D1563641

Finite Geometries, Buildings, and Related Topics

Finite Geometries, Buildings, and Related Topics

Edited by

William M. Kantor
University of Oregon

Robert A. Liebler
Colorado State University

Stanley E. Payne
University of Colorado at Denver

Ernest E. Shult
Kansas State University

CLARENDON PRESS · OXFORD
1990

Oxford University Press, Walton Street, Oxford OX2 6DP
Oxford New York Toronto
Delhi Bombay Calcutta Madras Karachi
Petaling Jaya Singapore Hong Kong Tokyo
Nairobi Dar es Salaam Cape Town
Melbourne Auckland
and associated companies in
Berlin Ibadan

Oxford is a trade mark of Oxford University Press

Published in the United States
by Oxford University Press, New York

British Library Cataloguing in Publication Data
Data available

Library of Congress Cataloging in Publication Data
Data available
[ISBN 0 19 853214 8]

Typeset by the author using TEX

Printed and bound in Great Britain by
Bookcraft Ltd, Midsomer Norton, Avon

Preface

The July 17-23, 1988, conference on Buildings and Related Geometries took place at Pingree Park, Colorado, in a scenic alpine cirque three miles from Rocky Mountain National Park. It brought together experts and interested mathematicians from around the world. The scientific program centered around invited expository lectures, and also included short research announcements and informal discussions.

This volume is intended to provide wider access to the expository portion of that conference. The editors hope that the snapshot of the subject that it provides will further stimulate interest in this area of mathematics. An earlier snapshot can be found in "Buildings and the Geometry of Diagrams", Springer Lecture Notes 1181, the proceedings of a related 1984 conference in Como, Italy. Since that conference, interest in buildings has accelerated while, at the same time, the subject has become more accessible to the mathematical community due to the recent publication of two introductory textbooks on the area (cf. p. 1 of these Proceedings).

The contributions to these proceedings have been ordered so as to begin with an introduction to various aspects of the theory, followed by papers gradually spreading out to more wide-ranging areas of geometry. In addition to the invited papers delivered at Pingree Park we have included work of J. Tits, who, because of illness, was unable to be present at the conference.

This book has been typeset in TEX. The editors acknowledge the extraordinary cooperation in this task of numerous contributors as well as the following TEX typists and formatters: Patricia Garcia (Colorado State University), Reta McDermott (Kansas State University), Mary Nickerson (University of Colorado at Denver) and W. Kantor (University of Oregon). Brad Shelton provided significant TEXnical advice, and Richard M. Koch programmed the diagram on p. 8. Martin Gilchrist, mathematics editor of Oxford University Press, has been very helpful and supportive throughout the preparation of this volume.

On behalf of the participants we thank the National Science Foundation, without whose support the conference would not have occurred. Valuable additional support came from Colorado State University and the University of Colorado at Denver.

W.M.K.
R.A.L.
S.E.P.
E.E.S.

Contents

List of Participants ix

List of Talks xi

Classification and Construction of Buildings 1
Mark Ronan

Spheres of Radius 2 in Triangle Buildings. I 17
Jacques Tits

A Census of Finite Generalized Quadrangles 29
Stanley E. Payne

Finite Geometries via Algebraic Affine Buildings 37
William M. Kantor

Groups Acting Transitively on Locally Finite Classical Tits Chamber Systems 45
Thomas Meixner

Representation Theory of Chambersystems of Affine Type 67
Udo Ott

The Geometry of Trilinear Forms 75
Michael Aschbacher

Local Recognition of Graphs, Buildings, and Related Geometries 85
Arjeh M. Cohen

Generalized Polygons and s-Transitive Graphs 95
Richard Weiss

Character Tables of Commutative Association Schemes 105
Eiichi Bannai

Codes and Curves 129
J. W. P. Hirschfeld

Solution of a Classical Problem on Finite Inversive Planes 145
J. A. Thas

List of Participants

NAME	AFFILIATION
Michael Adams	Colorado State University
Michael Aschbacher	California Institute of Technology
Catherine Baker	Mt. Allison University, Sackville, New Brunswick
Eiichi Bannai	Ohio State University
Lynn Batten	University of Manitoba
Curtis D. Bennett	University of Chicago
Aart Blokhuis	Technische Universiteit Eindhoven, The Netherlands
John van Bon	C. W. I., Amsterdam, The Netherlands
Julia M. N. Brown	York University, Downsview, Ontario
Xu-Ming Chen	Wayne State University
Andrew Chermak	Kansas State University
William Cherowitzo	University of Colorado at Denver
Arjeh M. Cohen	C. W. I., Amsterdam, The Netherlands
Hans Cuypers	Rijksuniversiteit Utrecht, The Netherlands
M. J. de Resmini	Università di Roma "La Sapienza", Italy
Gary L. Ebert	University of Delaware
J. C. Fisher	University of Regina, Saskatchewan
Daniel Frohardt	Wayne State University
Dina Ghinelli	Università di Roma "La Sapienza", Italy
Theo Grundhöfer	Universität Tübingen, West Germany
Diane Herrmann	University of Chicago
James Hirschfeld	University of Sussex, Falmer, Brighton, England
Sylvia Hobart	University of Wyoming
Tatsuro Ito	Joetsu University of Education, Joetsu, Niigata, Japan
Vikram Jha	Glasgow College, Scotland
Peter Johnson	University of Illinois at Chicago Circle
William M. Kantor	University of Oregon

Peter Kleidman	California Institute of Technology
Douglas Leonard	Auburn University
Robert A. Liebler	Colorado State University
Guglielmo Lunardon	Universitá di Napoli, Italy
Kay Magaard	California Institute of Technology
Thomas Meixner	Universität Giessen, West Germany
W. Mielants	Rijksuniversiteit-Gent, Belgium
G. Eric Moorhouse	University of Oregon
Akihiro Munimasa	Ohio State University
Theodore G. Ostrom	Washington State University
Udo Ott	Technische Universität Braunschweig, West Germany
Antonio Pasini	Universitá di Napoli, Italy
Stanley E. Payne	University of Colorado at Denver
Tim Penttila	University of New South Wales, Sydney, Australia
Laurel Rogers	University of Colorado at Colorado Springs
Mark Ronan	University of Illinois at Chicago Circle
Leanne Rylands	University of Sydney, Sydney, Australia
Yoav Segev	Michigan State University
Ernest E. Shult	Kansas State University
Sung-Yell Song	Ohio State University
Richard M. Stafford	National Security Agency
D. E. Taylor	University of Sydney, Sydney, Australia
Paul Terwilliger	University of Wisconsin
J. A. Thas	Rijksuniversiteit-Gent, Belgium
Hendrik Van Maldeghem	Rijksuniversiteit-Gent, Belgium
Diane Vuignier	University of Winnipeg
Richard Weiss	Tufts University
Frederick W. Wilke	University of Missouri at St. Louis
Satoshi Yoshiara	University of Illinois at Chicago Circle

List of Talks

M. Aschbacher	The geometry of trilinear forms
E. Bannai	Character tables of commutative association schemes
J. van Bon	Distance transitive graphs and linear groups
X. Chen	Groups that generate skew translation generalized quadrangles
A. M. Cohen	Local recognition of graphs, buildings, and related geometries
H. Cuypers	Generalized Fischer spaces
D. Ghinelli	Flag transitive extensions of C_n geometries
T. Grundhöfer	Topological polygons and affine buildings of rank 3
J. W. P. Hirschfeld	Codes and curves
T. Ito	t-Designs in coset geometries
V. Jha	On $GL(2, q)$ invariant spreads of order q^3
P. Johnson	Embedding geometries into projective spaces
W. M. Kantor	Affine buildings and their finite images
P. Kleidman	Ovoids
G. Lunardon	A result on C_n-geometries
K. Magaard	A structure theorem for $F_4(q)$ from the Jordan algebra point of view
T. Meixner	Groups acting on locally finite Tits chamber systems
W. Mielants	Algebraic curves and A_2-buildings
U. Ott	Representation theory of chamber systems of affine type
A. Pasini	Locally C_n geometries whose planes are affine
S. E. Payne	A census of finite generalized quadrangles
T. Penttila	Incidence matrices of diagram geometries
M. J. de Resmini	On the Johnson-Walker plane
M. Ronan	What is a building?
M. Ronan	Moufang buildings and their construction
L. Rylands	Representations and the split buildings of a group with a BN-pair
Y. Segev	On the uniqueness of the Monster

P. Terwilliger .... A generalization of the Bose-Mesner algebra of an association scheme

J. A. Thas .... Recent results on flocks, maximal exterior sets and inversive planes

J. A. Thas .... Solution of a classical problem on finite inversive planes

H. Van Maldeghem ... Topological polygons and affine buildings of rank 3

R. Weiss .... s-Transitive graphs and certain sporadic groups

S. Yoshiara .... On minimal weights of Steinberg codes

Classification and Construction of Buildings

Mark Ronan*

The purpose of this article is to give some feeling for what a building is, to discuss their classification, their construction, and something of their local structure. For more details two recent books, [Br] and [R2], are available, both published in 1989, plus a survey article [R3] which contains an introductory treatment of the subject.

As is well known, the theory of buildings was introduced, and has been largely developed, by Jacques Tits. Its original purpose was to provide an understanding of the analogues of the simple Lie groups, and particularly the exceptional groups, over an arbitrary field, and in this respect it has been brilliantly successful. The theory of spherical buildings and their classification [T2] was the result, but things by no means stop there. Work of Iwahori and Matsumoto [IM] gave us affine buildings; the general theory of affine buildings was then developed by Bruhat and Tits [BT], and a classification was accomplished by Tits [T7]. Later work, by Moody and Teo [MT], produced buildings for groups attached to Kac-Moody data, and these appear to be very interesting objects. Of course one should remember that the main reason for studying buildings is to use them for proving results about groups. In this respect affine buildings have been particularly important, and some of their applications are discussed in [R3].

This is a good place to stop and get down to basics.

1. Some Definitions

Before defining buildings, we start with Coxeter groups. Let I be a set and to each $i, j \in I$ let m_{ij} be a positive integer or ∞; we suppose $m_{ii} = 1$ and $m_{ij} = m_{ji} \geq 2$ if $i \neq j$. This data (m_{ij}) will be denoted M. The *Coxeter group* of type M is then a group W given by generators and relations as follows:

$$W = \langle r_i \mid (r_i r_j)^{m_{ij}} = 1 \text{ wherever } m_{ij} \neq \infty \rangle.$$

In particular $r_i^2 = 1$. This group may or may not be finite; for example if $I = \{i, j\}$ then W is the dihedral group of order $2m_{ij}$, which is finite, except of course when $m_{ij} = \infty$.

---

* Partially supported by the National Science Foundation.

It is convenient to encode the data M into a *diagram*. There is one node of the diagram for each $i \in I$, and an edge (or no edge) according to the following rule.

$i \quad j$

$\bullet \quad \bullet$ no edge if $m_{ij} = 2$

$\bullet$—$\bullet$ if $m_{ij} = 3$

$\bullet$═$\bullet$ if $m_{ij} = 4$

$\bullet \overset{m}{—} \bullet$ if $m_{ij} = m > 4$

If the diagram has several connected components then W is isomorphic to a direct product of Coxeter groups, one for each component of the diagram. To determine the finite Coxeter groups it therefore suffices to consider connected diagrams.

Theorem 1: *A finite Coxeter group is a direct product of Coxeter groups from the following list.*

	Diagram	$\lvert W\rvert$	Shape of W (*Atlas Notation*)
A_n		$(n+1)!$	S_{n+1}
C_n		$2^n n!$	$2^n S_n$
D_n		$2^{n-1}n!$	$2^{n-1}S_n$
E_6		$2^7 3^4 5$	$0_6^-(2).2$
E_7		$2^{10}3^4 5.7$	$2 \times 0_7(2)$
E_8		$2^{14}3^5 5^2 7$	$2.0_8^+(2).2$
F_4		1152	$2^3 : S_4 : S_3$
H_3	 5	120	$2 \times A_5$
H_4	5	$(120)^2$	$2A_5 : (2 \times A_5)$
$G_2(m)$	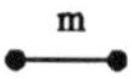 m	$2m$	D_{2m}

The subscript equals the number of nodes of the diagram; it is called the *rank* of W. In the A_n case, for example, the generators $r_1, r_2, \ldots, r_n$ can be taken to be the transpositions $(12), (23), \ldots, (n\ n+1)$ of S_{n+1}.

A finite Coxeter group is said to be of *spherical type* because it is the symmetry group of a triangulation of the $(n-1)$-sphere S^{n-1}, where n is the rank of W. This triangulation of S^{n-1} is the *Coxeter complex* for the group in question.

More generally, a Coxeter group of rank n acts non-trivially on $\mathbf{R}^n$, (see [T1] and [B]; also [R2] Chapter 2); it is therefore a non-trivial group, a fact which is not obvious from the presentation. In the spherical case this action is discrete, and a fundamental domain is called a *Weyl chamber*. It is a simplicial cone and intersects S^{n-1} in an $(n-1)$-simplex, called simply a *chamber*. The simplexes of dimension $(n-2)$ are called *panels*, and two chambers which have a panel in common are called *adjacent*. Notice that each panel is a face of exactly two chambers.

A spherical building of rank n and type M can now be described roughly in the following way. It is a simplicial complex of dimension $n-1$ in which any two chambers (i.e., $(n-1)$-simplexes) lie in a common subcomplex isomorphic to the Coxeter complex of type M, where M is a diagram for a finite Coxeter group. These subcomplexes are called *apartments*, and any two apartments which are not disjoint must intersect in a convex subset (in particular they are not just isomorphic, but there is an isomorphism from one to the other fixing their intersection).

Chamber Systems. To give a precise definition of a building, I prefer to treat buildings as "chamber systems" rather than simplicial complexes, because in the non-spherical case it is not always convenient to represent the chambers as simplexes. A *chamber system* (over I) is a set of objects called *chambers* together with, for each $i \in I$, an equivalence relation called *i-adjacency*. For example, in the case of a spherical building of rank n, there are n different types of panels and each chamber has one of each type; two chambers are then i-adjacent precisely when they share a panel of type i. Each i-adjacency class in a chamber system is called a *panel* of type i, and we think of panels as being faces of codimension 1.

Here is an example. Let W be a Coxeter group, and treat it as a chamber system by taking the elements $w \in W$ to be chambers, and the i-adjacency classes to be the left cosets of the subgroup of order 2 generated by r_i. In other words two chambers w and w' are i-adjacent if and only if $w' = w$ or wr_i. When W is finite we can represent chambers as $(n-1)$-simplexes on the sphere S^{n-1}, as mentioned earlier.

As an example in which it is not convenient to take chambers as simplexes, consider the diagram $M = \bullet\overset{\infty}{\text{—}}\bullet \quad \bullet\overset{\infty}{\text{—}}\bullet$. This is a rank 4 Coxeter group, isomorphic to the direct product of two infinite dihedral groups. It acts discretely on $\mathbf{R}^2$, preserving a square lattice.

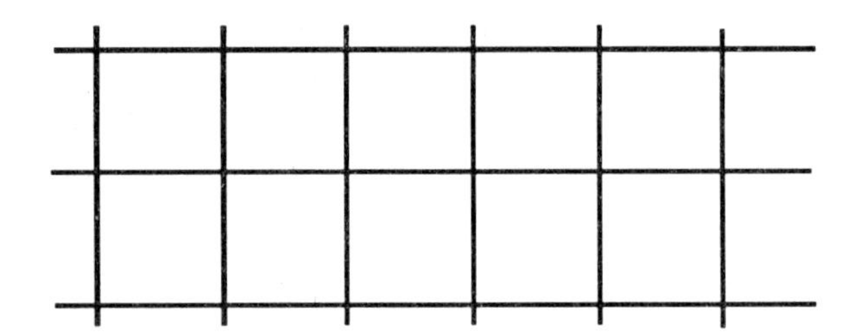

The natural Coxeter complex to take is this tiling of $\mathbf{R}^2$ by squares; each chamber is a square, and the panels are edges. As expected, each chamber has 4 panels (4 being the rank of W).

Residues and Galleries. A *gallery* in a chamber system is a sequence $(c_0, c_1, \ldots, c_\ell)$ of chambers each of which is adjacent to and distinct from the next; if c_{t-1} is i_t-adjacent to c_t we say the gallery has *type* $i_1 \ldots i_\ell$. A chamber system is called *connected* if any two chambers can be joined by a gallery.

Consider now a subset $J \subset I$. Take a chamber x and let R be the set of all chambers which can be joined to x by galleries whose types use only elements of J. Then R is a connected chamber system over J; it is called a *residue* of *type* J, or simply a J-*residue*.

Example: Treating W as a chamber system, a gallery from 1 to w is of the form $(1, r_{i_1}, r_{i_1}r_{i_2}, \ldots, r_{i_1} \ldots r_{i_\ell})$ where $w = r_{i_1} \ldots r_{i_\ell}$. If this gallery is of shortest possible length from 1 to w, then we say that $r_{i_1} \ldots r_{i_\ell}$ is a *reduced decomposition* for w, and the *word* $i_1 \ldots i_\ell$ is *reduced*. If W_J is the subgroup of W generated by all r_j for $j \in J$, then the chambers in a J-residue are precisely the elements in a left coset wW_J. It can be shown that W_J is a Coxeter group of type M_J, where $M_J = (m_{ij})$ for $i, j \in J$; in other words it inherits no further relations from the r_i for $i \notin J$.

Example: In the tiling of $\mathbf{R}^2$ by squares (above), each horizontal, or vertical, set of squares forms a residue whose types correspond to a connected component of the diagram. On the other hand a rank 2 residue of type $A_1 \times A_1$ (two disjoint nodes of the diagram) comprises the four squares having a vertex in common.

A Definition of Buildings. Let W, M, I be as before. A *building* Δ *of type* M is a chamber system over I, having a distance function d taking values in the Coxeter group W

$$d : \Delta \times \Delta \to W$$

and satisfying the following.

If $w = r_{i_1} \dots r_{i_\ell}$ is a reduced decomposition then $d(x, y) = w$ if and only if there is a gallery of type $i_1 \dots i_\ell$ from x to y.

This implies for example that $d(x, y) = r_i$ precisely when x and y are distinct and i-adjacent. Furthermore an easy induction argument shows that if a gallery of reduced type $i_1 \dots i_\ell$ exists then it is unique. It is then straightforward to show that if x is any chamber, and R any i-adjacency class, then there is a unique chamber of R nearest to x. In fact a much stronger result holds, namely if R is any residue then there is a unique chamber nearest to x – see e.g. [R2] p. 33.

We mentioned earlier that if $J \subset I$, then W_J is a Coxeter group of type M_J. Here is a stronger result whose proof is straightforward.

Theorem 2: *For a building of type M, each J-residue is a building of type M_J.*

Apartments. An *apartment* in a building Δ is a subcomplex A isomorphic to W (in other words there is a bijection $\varphi : W \to A$ such that $\varphi(w)$ and $\varphi(w')$ are i-adjacent in Δ if and only if w and w' are i-adjacent in W).

Theorem 3: *Any two chambers lie in a common apartment.*

This is an immediate corollary of a much stronger result, namely that if $X \subset W$ is any subset and $\varphi : X \to \Delta$ an isometry (i.e., preserving the distance $d(\ ,\ ) \in W$), then φ extends to an isometry (i.e., isomorphism) from W into Δ. This stronger result was proved in [T6] (3.7.4) – see also [R2] (3.5).

Digression on the Spherical Case. When Δ is of spherical type (i.e., when W is finite), the apartments are spheres (we are thinking here of Δ as a simplicial complex). In this case W has a unique element w_0 of longest length, namely that antipodal to 1. Choose a chamber $x \in \Delta$. For each chamber $y \in \Delta$ such that $d(x, y) = w_0$ (in which case we say x and y are *opposite*), there is a unique apartment containing x and y. These apartments form a basis for the top homology $H_{n-1}(\Delta)$, a fact which follows from the stronger result that Δ has the homotopy type of a bouquet of spheres $\vee S^{n-1}$, the number of spheres in the bouquet being precisely equal to the number of chambers y opposite x. To prove this one shows that after removing all those chambers y, the remaining complex is contractible; putting y back in again gives a sphere because its boundary gets identified to a point. A proof is given in [R2] Appendix 4.

Example of a Building. Let V be a vector space of dimension $n+1$ over a field k (not necessarily commutative), and let $I = \{1, \dots, n\}$. The $A_n(k)$

building has as its chambers all maximal flags of V;

$$0 \subset V_1 \subset \ldots \subset V_n \subset V.$$

Two chambers $0 \subset V_1 \subset \ldots \subset V_n \subset V$ and $0 \subset V_1' \subset \ldots \subset V_n' \subset V$ are i-adjacent precisely when $V_j = V_j'$ for all $j \neq i$. It is not obvious from our definition that this is a building, and indeed I shall not prove it. The apartments however are easy to obtain. Take a decomposition of V as a direct sum of 1-spaces $L_1 \oplus \ldots \oplus L_{n+1}$. These 1-spaces generate $(n+1)!$ maximal flags, namely:

$$0 \subset L_{\sigma(1)} \subset L_{\sigma(1)} \oplus L_{\sigma(2)} \subset \ldots \subset V$$

as σ ranges over the symmetric group S_{n+1}. This group S_{n+1} is isomorphic to W and acts in the obvious way on the apartment we have just obtained. Geometrically speaking the apartment is isomorphic to the barycentric subdivision of the boundary of an n-simplex.

2. Classification Theorems

For classification purposes one need only consider buildings whose diagrams are connected. In fact if $M_1, \ldots, M_r$ represent the connected components of M then one has the following theorem.

Theorem 4: *Every building of type M is a direct product $\Delta_1 \times \ldots \times \Delta_r$ where Δ_t is a building of type M_t.*

(By a direct product one means the following. If Δ_1 and Δ_2 are chamber systems over I_1 and I_2, then $\Delta_1 \times \Delta_2$ is a chamber system over $I_1 \coprod I_2$ whose chambers are all pairs (x_1, x_2) where $x_t \in \Delta_t$; if $i \in I_1$ then (x_1, x_2) is i-adjacent to (y_1, y_2) when x_1 is i-adjacent to y_1 and $x_2 = y_2$; similarly for $i \in I_2$.)

The Spherical Case. Tits [T2] has given a complete classification of (thick) spherical buildings of rank ≥ 3 with a connected diagram (thick means at least three chambers per panel). Here are some of the details (see also [R2] Chapter 8). For the diagrams, refer to Theorem 1.

Theorem 5: *The thick spherical buildings with a connected diagram of rank ≥ 3 are given in the following list.*

Diagram	**Buildings**
A_n	one for each field (not necessarily commutative)
D_n $(n \geq 4), E_6, E_7, E_8$	one for each commutative field
C_3 (with Cayley planes)	one for each Cayley division algebra
C_3 (with Desarguesian planes)	each such building arises from a pseudo-quadratic or sesquilinear form of Witt index 3 on a vector space over a field (not necessarily commutative)
C_n $(n \geq 4)$	as for C_3 (with Desarguesian planes), but with Witt index n.
F_4	there are five types, corresponding to: commutative fields, Cayley division algebras, quaternion algebras, certain "unitary" forms of Witt index 3 on a 6-space, and pairs (k, K) of imperfect fields of characteristic 2 with $k^2 \subset K \subset k$.
H_3, H_4	there are no such (thick) buildings

Notes: The A_n result is the classical theorem on projective space of dimension ≥ 3. The other results use a powerful theorem in Chapter 4 of [T2] which provides isomorphisms and automorphisms. Using this, the D_n, E_6, E_7, E_8 result is relatively straightforward. Existence for E_6, E_7, E_8 is obtained in [T2] by invoking the existence of algebraic groups of these types, though later in [RT] it is shown how to obtain these by an explicit construction – see also the last section of the present article. The C_n result depends essentially on C_3, which is accomplished in Chapters 7, 8, and 9 of [T2], and the F_4 result follows from C_3 and some further arguments in Chapter 10 of [T2]. The H_3 and H_4 result is contained in [T5], and is due to the fact that there are no "Moufang generalized 5-gons" – see below.

The Rank 2 Spherical Case. Here a classification is not possible. Rank 2 spherical buildings, also called *generalized m-gons* ($m = m_{ij}$), are equivalent to bipartite graphs of diameter m and girth $2m$ in which each vertex (panel) lies on at least two edges (chambers) – the *girth* is the length of a shortest circuit. For $m = 2$ this is usually called a complete bipartite graph. For $m \geq 3$ they are more interesting, and a generalized 3-gon is nothing other than the flag-graph of a projective plane E (i.e., the chambers are the flags of E, and the vertices are the points and lines of E). Illustrated below is the flag-graph of a plane of order 2; each circuit of length 6 in this graph is an apartment – there are 28 of them.

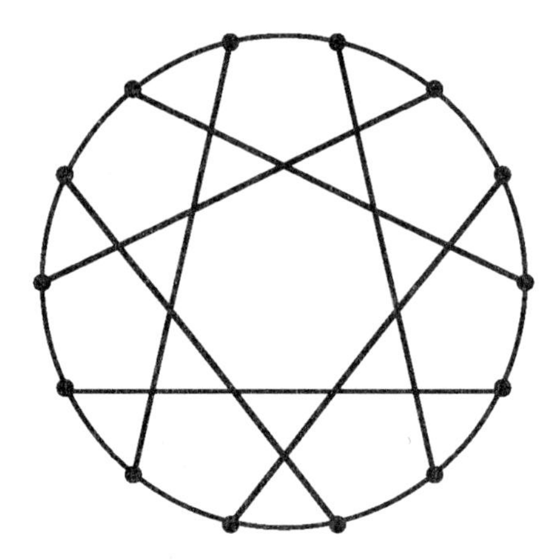

It is of course well-known that one cannot classify all projective planes unless extra conditions are imposed, such as the existence of certain automorphisms. The same is true for other generalized m-gons. Indeed it is possible to construct "free generalized m-gons" for any m (see [T5]), in a spirit similar to the construction of free projective planes. There is however an important class of generalized m-gons called *Moufang m-gons*, admitting large automorphism groups, for which there is an almost complete classification (see [T3] and the Addenda in [T2] – 1986 edition); in particular (thick) Moufang m-gons exist only for $m = 3, 4, 6$, or 8, a result due to Tits [T4] and Weiss [W].

It is also worth mentioning the Feit-Higman theorem on finite generalized m-gons.

Theorem 6 (Feit-Higman [FH]): *Finite generalized m-gons with at least three edges per vertex exist only for* $m = 2, 3, 4, 6$, *or* 8.

Examples of Moufang generalized m-gons, which can be either finite or infinite exist for $m = 3, 4, 6$, or 8 – see [T3], and also [R2] Appendix 2.

3. Affine Buildings and their Classification

A building is called *affine* (or *Euclidean*) if a typical apartment can be realized as a tiling of Euclidean space. For example the tiling of $\mathbf{R}^2$ by squares, which we saw earlier, is a rank 4 affine Coxeter complex; it is the direct product of two rank 2 ones, each with diagram $\overset{\infty}{\bullet\!\!-\!\!-\!\!\bullet}$.

Example: A rank 2 affine building necessarily has diagram $\overset{\infty}{\bullet\!\!-\!\!\!-\!\!\bullet}$. An apartment can be realized as the real line with integer points as panels, and unit intervals as chambers. Such buildings are equivalent to trees in which each vertex lies on at least two edges (they are, so to speak, generalized ∞-gons).

The connected affine diagrams are those for which the tiling of Euclidean space is a tiling by simplexes (as we have seen, the tiling by squares has a disconnected diagram). In this case each chamber of the building can be regarded as a Euclidean simplex with dihedral angles π/m_{ij}. Here is a list of all connected affine diagrams.

DIAGRAM	TYPE
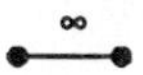 ∞	$\widetilde{A}_1$
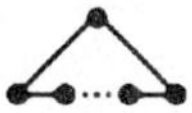	$\widetilde{A}_n,\ n \geq 2$
	$\widetilde{B}_n,\ n \geq 3$
	$\widetilde{C}_n,\ n \geq 2$
	$\widetilde{D}_n,\ n \geq 4$
	$\widetilde{E}_6$
	$\widetilde{E}_7$
	$\widetilde{E}_8$
	$\widetilde{F}_4$
 6	$\widetilde{G}_2$

Example: The $\widetilde{A}_n(K, v)$ building.

Let K be a field with a discrete valuation v. For instance the rationals $\mathbf{Q}$ with the p-adic valuation $v\left(\frac{a}{b}\, p^n\right) = n$ where p is a prime and $p \nmid a, b$; or indeed the p-adic numbers $\mathbf{Q}_p$ which is the completion of $\mathbf{Q}$ with respect to v. Another such example is the power series field $k((t))$, which is also complete. Now let $\mathcal{O} = \{a \in K \mid v(a) \geq 0\}$ be the valuation ring of K (e.g. $\mathbf{Z}_p \subset \mathbf{Q}_p$ or $k[[t]] \subset k((t))$), let V be a vector space over K, and let $L \subset V$ be an $\mathcal{O}$-lattice spanning V. Set two such $\mathcal{O}$-lattices L and L' equivalent if $L' = aL$ for some $a \in K$, and let x_L denote the equivalence class. The x_L are the vertices of the $\widetilde{A}_n(K, v)$ building Δ.

To determine the other simplexes, and indeed the chambers, of Δ, let $\pi \in \mathcal{O}$ be a "uniformizer" (a generator of the unique maximal ideal $\{a \in \mathcal{O} \mid v(a) > 0\}$). The quotient $k = \mathcal{O}/\pi\mathcal{O}$ is called the *residue field*; for instance when $K = \mathbf{Q}_p$ then k is the finite field with p elements. The simplexes are given by sets of vertices $x_{L_1}, \ldots, x_{L_m}$ where $L_1, \ldots, L_m$ can be chosen so that

$$L_1 \supset L_2 \supset \ldots \supset L_m \supset \pi L_1.$$

Given L_1 this corresponds to a flag in the k-vector space $L_1/\pi L_1$. As a consequence the largest possible m is n, where $n = \dim_k L_1/\pi L_1 = \dim_K V$, and Δ has dimension $n-1$ as a simplicial complex; each $(n-1)$-simplex is a chamber, and Δ has rank n as a building. Furthermore, the link of the vertex x_{L_1} is isomorphic to the $A_{n-1}(K)$ building, already described in section 1. When $n = 2$, Δ is a tree.

The Spherical Building at Infinity. Every affine building Δ of rank n has a spherical building Δ^∞ "at infinity", of rank $n-1$. For example when Δ is a tree, Δ^∞ is its set of ends.

To understand roughly how to get Δ^∞, take a single apartment A of Δ. It is a tiling of $\mathbf{R}^n$ and "at infinity" it induces a tiling A^∞ of the celestial sphere S^{n-1}. This is a Coxeter complex of spherical type, for a finite Coxeter group W_0, and is an apartment of Δ^∞. The affine Coxeter group W (acting on A) is a semi-direct product $\mathbf{Z}^n \rtimes W_0$ where the normal subgroup $\mathbf{Z}^n$ is a group of translations. Each possible complement W_0 is the stabilizer of a vertex of A, called a *special* vertex.

As far as the classification of affine buildings with a connected diagram is concerned, Δ^∞ is very important, because when Δ has dimension 3 (i.e., rank ≥ 4), then Δ^∞ has rank ≥ 3 and is therefore "known". As an example, suppose Δ is of type $\widetilde{E}_6$; then Δ^∞ is the $E_6(K)$ building for some commutative field K. The group of this type acts on Δ^∞, and one can show it also acts on Δ itself. This action on Δ allows one to obtain a discrete valuation v of K, and in fact K is complete with respect to v (the question of non-complete fields will be discussed shortly).

Generally speaking, one has the following theorem.

Theorem 7: *Every affine building of dimension ≥ 3 (with a connected diagram) arises over a field having a discrete valuation.*

For more precise details see [T7] or [R2] Chapter 10.

Before leaving affine buildings, I want to say a little more about fields, such as $\mathbf{Q}$, which are not complete with respect to a valuation v. Let K be such a field, and $\widehat{K}$ its completion with respect to v. Then for a given group, such as E_6, both fields lead to the *same* affine building Δ. However only $E_6(\widehat{K})$ is transitive on the set of all apartments. For $E_6(K)$ one takes a suitable orbit $\mathcal{A}$ of apartments, and using only apartments $A \in \mathcal{A}$ one obtains a smaller building at infinity, $(\Delta, \mathcal{A})^\infty \subset \Delta^\infty$. This $(\Delta, \mathcal{A})^\infty$ is the spherical building for $E_6(K)$. For more details see [T7] or [R2] Chapter 10.

4. Other Buildings

The question of classifying buildings which are neither spherical nor affine is wide open. However, it is known that in some cases a classification is not possible. For example when all rank 3 residues are non-spherical, a construction given in [R1] shows that there is too much freedom in the structure, and one cannot hope for a classification. On the other hand if each rank 3 subdiagram is of spherical type, then things would appear to be much more restricted, though no classification is yet in sight.

As an example, consider the (hyperbolic) diagram  . The only known buildings with this diagram arise from Kac-Moody groups, and at present we have no idea whether there are any others.

By contrast consider the affine diagram  . Here again there are buildings arising from Kac-Moody groups, such as $SL_4(k((t)))$ – an affine Kac-Moody group over the field k. On the other hand there are also buildings of this type arising from groups such as $SL_4(\mathbf{Q}_p)$, which is not an affine Kac-Moody group. Whether or not the p-adic affine buildings have analogues in the hyperbolic case is completely unknown.

5. Coverings – A Local Approach

Let I and M be as before, and let C be a chamber system over I such that for each 2-set $\{i,j\} \subset I$, every $\{i,j\}$-residue is a generalized m_{ij}-gon (so C is "locally" a building). Tits introduces such objects in [T6], where the concept of a chamber system first appears, and calls them *chamber systems of type M*. Of course each residue of type J is a chamber system of type M_J, and we shall say it is of *spherical type* if M_J is.

Example: The following example N, of type C_3, is not a building. It was first discovered by Neumaier [N], and later rediscovered by Aschbacher (see [N] or [R2] Chapter 4 for more details). Let S be a set of seven *points*, and L the set of thirty-five 3-sets in S, which we call *lines*. There are 30 ways of choosing seven lines to form a projective plane of order 2. Under the action of Alt_7 these split into two orbits of size 15; take one orbit P and call its members *planes*.

A *chamber* is then a triple (point, line, plane) where the point belongs to the line, and the line belongs to the plane. Two chambers are 1-, 2- or 3-adjacent if they differ in at most a point, a line, or a plane respectively. It should be clear that a $\{1,2\}$-residue is a projective plane (i.e., generalized 3-gon), and a $\{1,3\}$-residue is a generalized digon. Checking that a $\{2,3\}$-residue is a generalized 4-gon is a question of looking at the planes and the lines on a given point (see e.g. [R2] Chapter 4, Exercise 15). There are several ways to see that N is not a building. For example there are 315 chambers whereas a C_3 building with $2+1$ chambers per panel must have more than 2^9 chambers (9 being the length of the longest word in the Coxeter group of type C_3).

Coverings. A map $\varphi: C \to D$ between two chamber systems is called a *covering* (or 2-covering) if it is surjective and restricts to an isomorphism on any rank 2 residue. As in topology, one constructs a connected *universal covering* $\widetilde{C} \xrightarrow{\pi} C$ for any connected chamber system C. It has the usual universal property that if $\overline{C} \xrightarrow{\varphi} C$ is any covering for which $\overline{C}$ is connected, then there is a covering $\widetilde{C} \xrightarrow{\alpha} \overline{C}$ making the following diagram commute

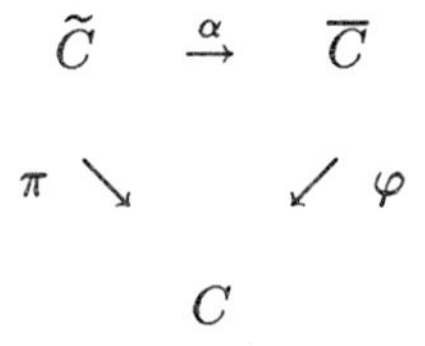

A chamber system which is its own universal cover is called *simply-connected*. All buildings are simply-connected. Furthermore the example N above is simply-connected; but that is rather exceptional as the following powerful theorem of Tits makes clear.

Theorem 8 (Tits [T6]): *If C is a chamber system of type M in which each rank 3 residue of spherical type is a building, then the universal cover of C is a building. In particular when C is simply-connected it is a building.*

6. Construction of Buildings

We conclude this article by briefly describing the construction of buildings given in [R1] and [RT].

The idea is to start with a given chamber c and work outwards. For each $w \in W$ one adjoins chambers x such that $d(c,x) = w$, working inductively on the length of w. If $w = w_i r_i$ where $\ell(w) = \ell(w_i) + 1$ for some $i \in I$, then x is adjacent to a unique chamber x_i for which $d(c, x_i) = w_i$; but the problem is that there may be several choices for $i \in I$. Thus x

might be required to be adjacent to several chambers $x_i, x_j, \dots$ which have already been constructed.

In [R1] it is assumed there are no rank 3 spherical subdiagrams, in which case this problem is overcome by ensuring that each rank 2 residue becomes a rank 2 building of the appropriate type. The resulting chamber system is simply-connected, and hence by Theorem 8 it is a building.

When there are rank 3 spherical subdiagrams, one must ensure that all residues of these types are buildings. To do this one uses "blueprints", introduced in [RT].

Blueprints. For each $i \in I$ let S_i be a chamber system over $\{i\}$ with distinguished element ∞_i; it will be a generic i-residue for the building of type M which we wish to construct. Now for each 2-set $\{i, j\} \subset I$ let S_{ij} be a generalized m_{ij}-gon with distinguished chamber ∞_{ij}.

We assume S_{ij} is *labelled*; that is to say, for each residue R of type $\{\alpha\}$ ($\alpha = i$ or j) there is a bijection

$$S_\alpha \xrightarrow{\varphi_\alpha} R$$

such that $\varphi_\alpha(\infty_\alpha)$ is the unique chamber of R nearest to ∞_{ij}. We think of S_i and S_j as a set of *labels*. In particular each chamber x which is i-adjacent to ∞_{ij} acquires a label $s_i \in S_i - \{\infty_i\}$. If x at distance $r_i r_j$ from ∞_{ij}, then there is an intermediary chamber x_j, and x corresponds to a sequence (s_i, s_j) of labels where x_j has the label s_i, and x has the label s_j. When x is opposite ∞_{ij} then there are two possible sequences $(s_i, s_j, \dots)$ and $(t_j, t_i, \dots)$, and to reconstruct S_{ij} it is necessary and sufficient to known when two such sequences give the same chamber. This data, for each S_{ij}, is called a *blueprint*.

The idea is then to start with a given chamber and work outwards, constructing each rank 2 residue according to the blueprint. When this gives a building we call the blueprint *realizable*. Of course not all blueprints are realizable; imagine one of type A_3 in which the A_2 residues are non-Desarguesian planes – there is no such building, so the blueprint cannot be realizable. However, one has the following theorem.

Theorem 9 ([RT] Theorem 1)*: *A blueprint is realizable if its restriction to each rank 3 spherical subdiagram is realizable.*

The question then is how to construct blueprints in such a way that they are obviously realizable for all rank 3 spherical subdiagrams. For disconnected subdiagrams it is sufficient that, whenever S_{ij} is a generalized

* There is an error in [RT] – the word spherical is missing from the statement of the theorem.

digon, (x_i, y_j) and (y_j, x_i) give the same chamber. For connected subdiagrams, namely A_3 and C_3 (we assume there is no H_3 subdiagram since there are no buildings of this type except the Coxeter complex itself), the problem is solved in [RT] by using root groups. The idea is to identify $S_i - \{\infty_i\}$ with a root group U_i.

Example: The Natural A_2 blueprint (for a Desarguesian plane). In this case we have two root groups U_i and U_j, both of which can be identified with the additive group of a field k. We identify $u \in k$ with $u_i \in U_i$ and $u_j \in U_j$. We can then say that two sequences (x_i, y_j, z_i) and (a_j, b_i, c_j) give the same chamber if and only if:

$$x = c, \quad z = a, \text{ and } y + b = xz.$$

Here xz means multiplication in the field k.

Now to construct an A_3 building we use this A_2 blueprint (for each A_2 subdiagram), and the $A_1 \times A_1$ blueprint mentioned above. The resulting A_3 blueprint is realizable when the two field structures induced on the middle node of the diagram (one from each A_2 residue) are opposite one another.

As an immediate application, consider a D_4 diagram. The three A_2 residues induce field structures on the middle node of the diagram, and these fields are mutually opposite. Thus the field is commutative in this case.

Finally suppose we wish to construct an E_8 building. Choose a commutative field (because of the D_4 subdiagram), and use the A_2 and $A_1 \times A_1$ blueprints as above. Theorem 9 does the rest. This is an amazingly simple existence proof for E_8 buildings. As a corollary, the result of Tits [T2] Chapter 4 on the existence of automorphisms gives an immediate existence proof for groups of type E_8 over an arbitrary commutative field.

There is one last point I want to mention. All spherical buildings can be obtained from blueprints, and more generally so can all buildings attached to Kac-Moody groups. However not all affine buildings arise in this way; those obtained from p-adic groups, such as $SL_n(\mathbf{Q}_p)$, do not conform to a blueprint. The question of constructing such buildings is an interesting open problem.

References:

[B] N. Bourbaki, Groupes et Algèbres de Lie, Ch. 4, 5 et 6. Hermann, Paris 1968; Masson, Paris 1981.

[Br] K. S. Brown, Buildings, Springer-Verlag 1989.

[BT] F. Bruhat et J. Tits, Groupes réductifs sur un corps local, I. Données radicielles valuées, *Publ. Math. I.H.E.S.*, **41**(1972), 5-252.

[FH] W. Feit and G. Higman, The nonexistence of certain generalized polygons, *J. Algebra,* **1**(1964), 114-131.

[IM] N. Iwahori and H. Matsumoto, On some Bruhat decomposition and the structure of the Hecke rings of $\mathfrak{p}$-adic Chevalley groups, *Publ. Math. I.H.E.S.,* **25**(1965), 5-48.

[MT] R. Moody and K. Teo, Tits Systems with crystallographic Weyl groups, *J. Algebra,* **21**(1972), 178-190.

[N] A. Neumaier, Some Sporadic Geometries related to $PG(3,2)$, *Arch. Math.,* **42**(1984), 89-96.

[R1] M. A. Ronan, A Construction of Buildings with no Rank 3 Residues of Spherical Type, in *Lecture Notes in Mathematics* **1181** (Buildings and the Geometry of Diagrams, Como 1984), Springer-Verlag 1986, 242-248.

[R2] M. A. Ronan, Lectures on Buildings, Academic Press (Perspectives in Mathematics 7), 1989.

[R3] M. A. Ronan, Buildings: Main Ideas and Applications, *Bulletin of the London Mathematical Society,* (to appear).

[RT] M. A. Ronan and J. Tits, Building Buildings, *Math. Annalen,* **278** (1987), 291-306.

[T1] J. Tits, Le problème des mots dans les groupes de Coxeter. *Ist Naz. Alta Mat., Symposia Math.,* **1**(1968), 175-185.

[T2] J. Tits, Buildings of Spherical Type and Finite BN-Pairs, *Lecture Notes in Mathematics,* **386**, Springer-Verlag (1974), (2nd edition 1986).

[T3] J. Tits, Classification of Buildings of Spherical Type and Moufang Polygons: A survey, in *Atti. Coll. Intern. Teorie Combinatorie,* Accad. Naz. Lincei, Rome, 1973 vol.1 (1976), 229-246.

[T4] J. Tits, Non-existence de certains polygones généralisés, I, II. *Inventiones Math.* **36**(1976), 275-284; **51**(1979), 267-269.

[T5] J. Tits, Endliche Spiegelungsgruppen, die als Weylgruppen auftreten, *Inventiones Math.,* **45**(1977), 283-295.

[T6] J. Tits, A Local Approach to Buildings, in The Geometric Vein (Coxeter Festschrift), Springer-Verlag, 1981, 519-547.

[T7] J. Tits, Immeubles de type affine, in *Lecture Notes in Mathematics,* **1181**, (Buildings and the Geometry of Diagrams, Como 1984), Springer-Verlag, 1986, 159-190.

[W] R. Weiss, The Nonexistence of certain Moufang Polygons, *Inventiones Math.,* **51**(1979), 261-266.

Spheres of Radius 2 in Triangle Buildings. I

Jacques Tits

1. Introduction

A *triangle building* is a building Δ of type $\tilde{A}_2$, i.e. whose Coxeter diagram is a triangle

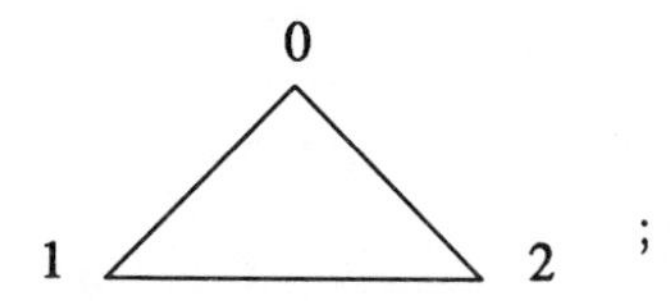

;

as on the picture, we shall always number 0, 1, 2 the three types of vertices of Δ. We denote by $\mathrm{Vert}\Delta$ (resp. $\mathrm{Vert}_i\Delta$) the set of all vertices (resp. all vertices of type i) of Δ, and always suppose Δ to be thick, in which case all 1-simplices are contained in the same number of 2-simplices (or chambers); if that number is $q+1$, we call q the *order* of Δ. If $v \in \mathrm{Vert}\Delta$, the full subcomplex $S_2(v)$ of Δ whose vertices are all vertices of Δ at distance ≤ 2 of v in the "graph of Δ" (the graph whose vertices and edges are the 0- and 1-simplices of Δ) is called the *sphere of radius* 2 with center v in Δ. The main purpose of this paper is to show that

the spheres of radius 2 *in triangle buildings of order* 2 *form exactly two isomorphism classes.*

We distinguish those two classes by means of an invariant taking its values in $\mathbf{Z}/2\mathbf{Z}$. Computing that invariant in the case of the four buildings obtained, after M. Ronan, by amalgamation of three Frobenius groups of order 21 (cf. [3], [5], [6]) enables us to show that those four buildings are pairwise nonisomorphic and that two of them are not classical, that is, are not isomorphic to affine buildings of groups of type A_2 over a locally compact local field (whereas the two others are known to be classical: cf. [1] and [2]). The 44 buildings of order 8 obtained by amalgamation of three Frobenius groups of order $9 \cdot 73$ (cf. [6], 3.2) can be approached by similar (but considerably more involved) methods, as I hope to show in a forthcoming paper.

In the last section of the paper, a rough lower bound is given for the number $N(q)$ of isomorphism classes of spheres of radius 2 in triangle buildings of given order q, when q is a prime power (but the method of proof extends with minor modifications to any q such that there exists a projective plane of order q). That bound (and, no doubt, the number $N(q)$ itself) increases very rapidly with q. For instance, whereas our main result says that $N(2) = 2$, it follows from Corollary 5 in 5.1 that $N(3) \geq 11950$ (probably a rather weak estimate), $N(4) \geq 1.6 \cdot 10^{69}$, $N(5) \geq 2.5 \cdot 10^{306}$, etc. Corollary 5′ of 5.2, a variant of the previous one, shows that the number of isomorphism classes of spheres of radius 2 remains very large (for $q \geq 3$) even when one considers only "triangle buildings over $\mathbf{F}_q$", as defined in 5.2.

This paper is extracted from personal notes of a course given at the Collège de France in 1984-1985. It has been pointed out to me by W. Kantor that, in the published summary of that course [6], the section relative to the results presented here does not make sense because of an important slip in the definitions on top of p. 103.

2. An alternative description of the spheres of radius 2

To start with, we recall a standard definition and fix some notation. If P and P' are two projective planes, dual to each other, a pair $(x, x') \in P \times P'$ is called a *flag* when x belongs to the line of P represented by x', in which case, we also say that x' is *incident* or *orthogonal* to x, and we write $x \perp x'$.

Let Δ be a triangle building, let ω be a vertex of Δ, say of type 0, and, for $i = 1$ or 2, let P_i be the set of all vertices of type i of Δ incident to ω. Thus, P_1 and P_2 are the sets of points of two projective planes, dual to each other. For $(x_1, x_2) \in P_1 \times P_2$ with $x_1 \perp x_2$, we denote by $C(x_1, x_2)$ or, indifferently, by $C(x_2, x_1)$ the set of all chambers of Δ containing the 1-simplex $\{x_1, x_2\}$ and different from $\{\omega, x_1, x_2\}$. For $x \in P_i$ ($i = 1$ or 2), the set of all vertices of type 0 of Δ incident to x has a structure of projective plane with a distinguished point ω; we denote it by Q_x. The lines of Q_x which contain ω are in one-to-one correspondence with the vertices $y \in P_{3-i}$ incident to x. For such a y, the corresponding line is in canonical bijection with the set $C(x, y) \cup \{\omega\}$. Thus, to (Δ, ω), we have associated a system

$$\mathcal{S} = (\omega; P_i\,(i \in \{1,2\})\,; C; Q_x\,(x \in P_1 \cup P_2))\,,$$

consisting of a point ω, two projective planes P_1, P_2, each the dual of the other, a function C attaching a set $C\,(x_1, x_2)$ to every flag $(x_1, x_2) \in P_1 \times P_2$ (with $x_1 \perp x_2$) and, for any $x \in P_i$, with $i \in \{1, 2\}$, a structure Q_x of projective plane on the set

$$\{\omega\} \cup \bigcup \{C(x, y) \mid y \in P_{3-i},\ y \perp x\}$$

such that the subsets $\{\omega\} \cup C(x, y)$, with y as above, are the lines of Q_x containing ω. Observe that *all data in $\mathcal{S}$ are provided by the sphere of radius 2 with center ω in Δ.*

Conversely,

any such system $\mathcal{S}$ determines a 2-dimensional complex S_2 which can be embedded as a sphere of radius 2 in a triangle building.

Indeed, the vertices of S_2 are all points and all lines of the projective planes Q_x (for $x \in P_1 \cup P_2$) with the following identifications:

the point ω is common to all planes Q_x;

for $(x_1, x_2) \in P_1 \times P_2$, $x_1 \perp x_2$, $x_1 \in P_1$ and $x_2 \in P_2$, the set $C(x_1, x_2)$ considered as a subset of Q_{x_1} is identified with the same set considered as a subset of Q_{x_2};

for $x \in P_i$ and $y \in P_{3-i}$ incident to x, the vertex representing the line $\{\omega\} \cup C(x, y)$ of Q_y does not depend on y: we simply denote it by x, thus embedding P_i in the set of vertices.

We further define an incidence relation between vertices as follows: two vertices are incident if and only if

they coincide,

or one of them, call it v, belongs to $P_1 \cup P_2$ and the other is a point or a line of Q_v,

or they are a point and a line incident to each other in some Q_x, with $x \in P_1 \cup P_2$.

Finally, the 1-simplices and the 2-simplices of S_2 are the pairs of distinct incident vertices and the triples of pairwise distinct and incident vertices. That the complex S_2 obtained in that way can be embedded as a sphere of radius 2 in a triangle building readily follows from Ronan's construction of such buildings (cf. [4]).

A system $\mathcal{S}$ as above will sometimes simply be called *a sphere*, for short. The cardinality of the set $C(x, y)$ does not depend on the flag $(x, y) \in P_1 \times P_2$ and is equal to the order of the planes P_1 and P_2; we call it the *order* of the sphere $\mathcal{S}$.

3. Spheres of order 2

Let $\mathcal{S}$ be a sphere of order 2. For each flag $(x, y) \in P_1 \times P_2$ (with $x \perp y$), let us choose a bijection $\lambda(x, y)$ of $\mathbf{Z}/2\mathbf{Z} = \{0, 1\}$ onto $C(x, y)$. Such a function λ will be called a *labelling* of the sphere $\mathcal{S}$. For $x \in P_i$, there are three points of P_{3-i} incident to x, say y, y', y''. The plane Q_x consists of the seven points

$$\omega, \lambda(x, y)(z), \lambda(x, y')(z), \lambda(x, y'')(z) \quad (z = 0, 1).$$

Clearly, for one and only one value of z the three points $\lambda(x,y)(z)$, $\lambda(x,y')(z)$,$\lambda(x,y'')(z)$ are collinear and that value, which we denote by $\xi_\lambda(x)$, or simply $\xi(x)$, determines the projective plane structure Q_x on the seven points in question: for instance, if $\xi(x) = 0$, the seven lines of Q_x are

$$\{\lambda(x,y)(0), \lambda(x,y')(0), \lambda(x,y'')(0)\},$$

$$\{\omega, \lambda(x,\sigma(y))(0), \lambda(x,\sigma(y))(1)\},$$

$$\{\lambda(x,\sigma(y))(0), \lambda(x,\sigma(y'))(1), \lambda(x,\sigma(y''))(1)\},$$

where σ runs over all cyclic permutations of (y, y', y'').

Lemma: *If we modify the labelling λ at exactly one place (x_0, y_0), that is, if we add 1 to $\lambda(x_0, y_0)$ and leave all other $\lambda(x,y)$ invariant, then all values of ξ remain unchanged, except for $\xi(x_0)$ and $\xi(y_0)$ to which 1 is added.*

This is obvious.

Theorem: *The sum $\xi(\mathcal{S}) = \sum \xi_\lambda(x)$ over all $x \in P_1 \cup P_2$ depends only on $\mathcal{S}$, not on the chosen labelling λ; it determines the isomorphism type of the sphere $\mathcal{S}$.*

By the Lemma, if one modifies the labelling λ at exactly one place, the sum $\sum \xi_\lambda(x)$ remains unchanged, hence the first assertion.

Let us now denote the 14 elements of $P_1 \cup P_2$ by x_j, with $j = 1, 2, \ldots,$ 14, the indexing being chosen in such a way that $x_1 \in P_1$, say, and that $x_j \perp x_{j+1}$ for all $j \leq 13$: it is well known that such an ordering of the points and lines of a projective plane of order 2 exists. Starting from a labelling λ of $\mathcal{S}$ and the corresponding functions $\xi = \xi_\lambda$, we define another labelling μ as follows:

if $j \in \{1, \ldots, 13\}$, we set

$$\mu(x_j, x_{j+1}) = \lambda(x_j, x_{j+1}) + \sum_{k=1}^{j} \xi(x_k)$$

(here, we extend the notation λ, μ by setting $\lambda(y,x) = \lambda(x,y)$ and $\mu(y,x) = \mu(x,y)$);

if $(x,y) \in P_1 \times P_2$ is a flag such that, for no value of $j \leq 13$ one has $(x,y) = (x_j, x_{j+1})$ or (x_{j+1}, x_j), then we take $\mu(x,y) = \mu(y,x) = \lambda(x,y)$.

From the Lemma, it now readily follows that

$$\xi_\mu(x_j) = 0 \quad \text{for all } j \leq 13,$$

and then, the first part of the theorem, already proved, implies:

$$\xi_\mu(x_{14}) = \xi(\mathcal{S}).$$

Therefore, $\xi(\mathcal{S})$ determines the function ξ_μ which, in turn, determines the projective plane structures Q_x, hence the isomorphism type of $\mathcal{S}$, q.e.d.

4. Computation of the invariant $\xi(\mathcal{S})$ for the buildings associated to amalgamated sums of three Frobenius groups of order 21

We first recall the definition of the buildings in question.

A Frobenius group of order 21 is nothing else but a nonabelian group of order 21. Such a group X has a normal subgroup K of order 7 and every element of $X-K$ generates a *Frobenius complement*, of order 3. All ordered pairs of Frobenius complements are equivalent under the full automorphism group of X, but the two nonneutral elements of any Frobenius complement play asymmetric roles: indeed, one and only one of them, which we shall call *squaring*, conjugates every element of K into its square.

We now consider three Frobenius groups of order 21, which we call X_i ($i = 0, 1, 2$), in each X_i, we choose two distinct Frobenius complements, denoted by Y_{ji} and Y_{ki}, where (i, j, k) is a permutation of $(1, 2, 3)$, and we form the amalgamated sum G of X_0, X_1, X_2, amalgamating Y_{01} with Y_{02}, Y_{12} with Y_{10} and Y_{20} with Y_{21}. It is known (cf. e.g. [7], Theorem 1) that the canonical maps $X_i \to G$ are injective. Thus, an alternative description of G is as follows: it is generated by three subgroups Y_i ($i = 0, 1, 2$) of order 3 (in the first description, $Y_i = Y_{ij} = Y_{ik}$), for any permutation (i, j, k) of $(0, 1, 2)$, Y_j and Y_k generate a group X_i of order 21 and defining relations for the three groups X_i constitute a complete set of defining relations for G.

There is no uniqueness for the situation because, when amalgamating Y_{ij} with Y_{ik}, we may identify the squaring element of Y_{ij} with the squaring element of Y_{ik} or with its inverse. We introduce an invariant $c_i \in \mathbf{Z}/2\mathbf{Z}$ equal to 0 in the first case and to 1 in the second case. When it is necessary to specify the three choices made for the amalgamation, we set $G = G(c_0, c_1, c_2)$. Clearly, if (c'_0, c'_1, c'_2) is a permutation of (c_0, c_1, c_2), the groups $G(c_0, c_1, c_2)$ and $G(c'_0, c'_1, c'_2)$ are isomorphic, thus the groups G we have defined belong to at most four isomorphism classes, namely the classes of $G(0,0,0)$, $G(0,0,1)$, $G(0,1,1)$ and $G(1,1,1)$. (In fact, it can be shown with the method of [8] that those four classes are indeed different.)

The complex whose set of vertices is the disjoint union of the three sets G/X_i and whose simplices are the sets of vertices with nonempty intersections (the vertices in question being regarded as subsets of G) is a triangle building (cf. [3] or [7], *loc. cit.*) which we denote by Δ or, more precisely, $\Delta(c_0, c_1, c_2)$. The vertices belonging to G/X_i will of course be given the type i. Observe that the chambers of Δ are (in canonical 1 - 1 correspondence with) the elements of G.

Proposition 1: *The invariant ξ of the spheres of radius 2 in $\Delta(c_0, c_1, c_2)$ with center a vertex of type i is equal to c_i.*

We assume, without loss of generality, that $i = 0$, denote by ω_j ($j = 0, 1, 2$) the vertex X_j of Δ, set $\omega = \omega_0$, consider the sphere $\mathcal{S}$ of radius 2 and center ω in Δ and use the notation of §2.

We can identify the flags in $P_1 \times P_2$ with the chambers containing ω (by $(x, y) \leftrightarrow \{\omega, x, y\}$), that is, with the elements of X_0. Then, the function C of §2 is given by

$$C(x) = x \cdot (Y_0 - \{1\}) \quad (x \in X_0).$$

For $j = 1, 2$, let u_j be the element of $Y_0 - \{1\}$ identified with the squaring element of Y_{0j}, and let λ_j be the labelling of $\mathcal{S}$ defined by

$$\lambda_j : xu_j \mapsto 0, \quad xu_j^{-1} \mapsto 1 \quad (x \in X_0).$$

The group X_0 acts on $\mathcal{S}$ (by left translations), permutes P_k transitively for $k = 1, 2$, and preserves λ_j. Therefore, ξ_{λ_j} is constant on P_k; we denote by ξ_{jk} its value on that set. According as $c_0 = 0$ or 1, we have $u_2 = u_1$, hence $\lambda_2 = \lambda_1$ and $\xi_{2k} = \xi_{1k}$, or $u_2 = u_1^{-1}$, hence $\lambda_2 = \lambda_1 + 1$ and $\xi_{2k} = \xi_{1k} + 1$. Thus, in all cases,

$$\xi_{2k} = \xi_{1k} + c_0. \tag{1}$$

We also have, by symmetry, $\xi_{12} = \xi_{21}$. Finally, $\xi(\mathcal{S})$ is the sum of all values of ξ on $P_1 \times P_2$, hence equal to

$$7\xi_{11} + 7\xi_{12} = \xi_{11} + \xi_{12} = \xi_{11} + \xi_{21} = c_0,$$

q.e.d.

Remark 1: For the proof, we have not needed to know the exact values of the ξ_{jk}, but they are easily determined. Indeed, set $u_1 = u$ and let v denote the canonical image of the squaring element of Y_{21} in Y_2. Since u and v are both squaring elements in X_1, their quotient $a = uv^{-1}$ belongs to the subgroup of order 7 of X_1. Therefore, we have

$$au = a^2v = va,$$

hence, for all n,

$$au^n = v^n a$$

and also

$$au^n X_0 = v^n a X_0 = v^n u v^{-1} X_0 = v^n u X_0.$$

Since $\{X_0, uX_0, u^2X_0\}$ is a line in the plane Q_{ω_1} (namely the line of all points of that plane incident to ω_2), its left translate by a, that is, in view of the above identity, the set $\{uX_0, vuX_0, v^2uX_0\}$, is also a line. With the identification

$$Q_{\omega_1} \leftrightarrow \{\omega\} \cup C(1) \cup C(v) \cup C\left(v^2\right)$$

of §2 (as before, the elements of X_0 also stand for flags in $P_1 \times P_2$), that same line can be written $\{u, vu, v^2u\}$. Its three elements are therefore labelled 0 in the labelling λ_1, and, by definition of ξ_{ij}, this implies that $\xi_{11} = 0$. By symmetry, $\xi_{22} = 0$. Finally, in view of (1), $\xi_{12} = \xi_{21} = c_0$.

Corollary 1: *If there exists a type-preserving isomorphism*

$$\Delta\left(c_0, c_1, c_2\right) \to \Delta\left(c_0', c_1', c_2'\right),$$

then $c_i = c_i'$ for all i. The four buildings $\Delta(0,0,0)$, $\Delta(0,0,1)$, $\Delta(0,1,1)$, $\Delta(1,1,1)$ are pairwise nonisomorphic (even when non-type-preserving isomorphisms are allowed).

Corollary 2: *If there exists a (non-type-preserving) automorphism of $\Delta(c_0, c_1, c_2)$ mapping the vertices of type j onto the vertices of type k, then $c_j = c_k$.*

Corollary 3: *If $\Delta(c_0, c_1, c_2)$ is isomorphic to the affine building $\Delta(K)$ of $SL_3(K)$, where K is a locally compact local field with residue field $\mathbf{F}_2$, then $c_0 = c_1 = c_2$.*

Indeed, if π is a uniformizing element of K, the matrix

$$\begin{pmatrix} 0 & 1 & 0 \\ 0 & 0 & 1 \\ \pi & 0 & 0 \end{pmatrix}$$

acts on $\Delta(K)$ and permutes the three types of vertices cyclically.

Remark 2: By results of P. Köhler, T. Meixner and M. Wester, $\Delta(1,1,1)$ and $\Delta(0,0,0)$ are respectively isomorphic to the affine buildings of $SL_3(\mathbf{Q}_2)$ (cf. [1]) and $SL_3(\mathbf{F}_2((t)))$ (cf. [2]).

Exercises. (i) Show that if K is a field endowed with a complete discrete valuation, if $\mathcal{O}$ is its valuation ring and π a uniformizing element, then the isomorphism type of the spheres of radius 2 in the affine building of $SL_3(K)$ depends only on the ring $\mathcal{O}/\pi^2\mathcal{O}$. [*Hint.* The spheres in question are, in some sense, the building of the matrix algebra $\mathbf{M}(\mathcal{O}/\pi^2\mathcal{O})$. Find the appropriate definition.]

(ii) From Exercise (i) and Remark 2, deduce that if K is a locally compact local field with residue field $\mathbf{F}_2$, the invariant ξ of the spheres of radius 2 of the affine building of $SL_3(K)$ is equal to 1 or 0 according as K is isomorphic to $\mathbf{Q}_2$ or not.

(iii) Prove the last assertion by direct computations.

5. A lower bound for the number of isomorphism classes of spheres of radius 2 in triangle buildings of given order

It will be convenient to consider only type-preserving isomorphisms but, clearly, if we find a lower bound for the number of type-preserving isomorphism classes, one half of that bound will be a bound for the number of full isomorphism classes. We shall also assume that all projective planes occurring are Desarguesian (and, of course, finite), which implies that the given order q is a prime power p^n. This hypothesis plays no essential role in the arguments but the formulas which one obtains are simpler and more explicit in that they would otherwise involve such parameters as the orders of the automorphism groups of the projective planes in question. Besides, if q is assumed to be a prime power, the lower bound which one obtains when considering only Desarguesian planes remains valid, *a fortiori*, without that restriction.

5.1. General buildings

Let P_1, P_2 be two Desarguesian projective planes of order q, dual to each other and for each flag $(a,b) \in P_1 \times P_2$, $a \perp b$, let there be given a set $C(a,b) = C(b,a)$ of cardinality q. We shall be interested in the set $\mathcal{T}$ of all spheres

$$\mathcal{S} = \big(\omega; P_1, P_2; C; \big(Q_x \mid x \in P_1 \cup P_2\big)\big)$$

(cf. §2), where the data P_1, P_2 and C are the given ones and the projective plane structures Q_x are allowed to vary. For $x \in P_i$ $(i = 1,2)$, let Σ_x be the group of all permutations of the set

$$\bigcup \{C(x,y) \mid y \in P_{3-i},\ y \perp x\}$$

preserving its partition into the subsets $C(x,y)$ and, for $\mathcal{S} \in \mathcal{T}$ as above, let $\Sigma_x^{\mathcal{S}}$ $(\subset \Sigma_x)$ be the group of all automorphisms of the projective plane Q_x which fix ω. The orders of those groups are

$$|\Sigma_x| = (q!)^{q+1}(q+1)!$$

and

$$\left|\Sigma_x^{\mathcal{S}}\right| = n \cdot q^3 \left(q^2 - 1\right)(q-1).$$

We set $\Sigma = \prod \Sigma_x$ and $\Sigma^{\mathcal{S}} = \prod \Sigma_x^{\mathcal{S}}$, where x runs over $P_1 \cup P_2$ and we denote by Σ_0 the direct product of the full permutation groups of the sets $C(x,y)$. This product can be regarded as a subgroup of Σ: if, for all $(a,b) \in P_1 \times P_2$, $a \perp b$, σ_{ab} is a permutation of $C(a,b)$, the element (σ_{ab}) of Σ_0 is identified with the element of Σ whose component in Σ_x stabilizes each $C(x,y)$ $(y \perp x)$ and induces σ_{xy} or σ_{yx} on that set. The order of the group Σ_0 is

$$|\Sigma_0| = (q!)^{(q+1)(q^2+q+1)}.$$

The group Σ acts on $\mathcal{T}$ in the obvious way: if $\sigma \in \Sigma$ and if $\mathcal{S} \in \mathcal{T}$ is as above, $\sigma\mathcal{S}$ is the sphere $(\omega; P_1, P_2; C; (Q'_x))$, where Q'_x is Q_x transformed by the Σ_x-component of σ. Because of the hypothesis made that the projective planes Q_x are all Desarguesian (and of order q), Σ is transitive on $\mathcal{T}$.

Proposition 2: *For $\mathcal{S} \in \mathcal{T}$ and $\sigma \in \Sigma$, there exists an isomorphism $\mathcal{S} \to \sigma\mathcal{S}$ fixing P_1 and P_2 if and only if $\sigma \in \Sigma_0 \cdot \Sigma^{\mathcal{S}}$.*

Just observe that an isomorphism $\sigma' : \mathcal{S} \to \sigma\mathcal{S}$ fixing P_i "is" an element of Σ_0, and that $\sigma'^{-1}\sigma \in \Sigma^{\mathcal{S}}$.

Let us say that two elements $\mathcal{S}$, $\mathcal{S}'$ of $\mathcal{T}$ are P_1-isomorphic if there exists an isomorphism of $\mathcal{S}$ onto $\mathcal{S}'$ fixing P_1.

Corollary 4: *The number of P_1-isomorphism classes in $\mathcal{T}$ is at least $|\Sigma| \cdot \left|\Sigma^{\mathcal{S}}\right|^{-1} \cdot |\Sigma_0|^{-1}$ (for any $\mathcal{S} \in \mathcal{T}$).*

The last expression is equal to

$$\begin{aligned}
&(q!)^{2(q+1)(q^2+q+1)} \left((q+1)!\right)^{2(q^2+q+1)} . \\
&\qquad \left(nq^3 \left(q^2-1\right)(q-1)\right)^{-2(q^2+q+1)} (q!)^{-(q+1)(q^2+q+1)} \\
&= q^{(q-3)(q^2+q+1)} (q-1)^{q^3-1} \left((q-2)!\right)^{(q+3)(q^2+q+1)} n^{-2(q^2+q+1)}.
\end{aligned}$$

Dividing by

$$|\mathrm{Aut}\ P_1| = nq^3(q-1)^2 \left(q^2+q+1\right)(q+1),$$

we get:

Corollary 5: *The number of isomorphism classes in $\mathcal{T}$ is $\geq$*

$$\begin{aligned}
&q^{(q-3)(q^2+q+1)-3} (q-1)^{q^3-3} \left((q-2)!\right)^{(q+3)(q^2+q+1)} . \\
&\qquad n^{-2(q^2+q+1)-1} \left(q^2+q+1\right)^{-1} (q+1)^{-1}.
\end{aligned}$$

This formula provides the estimates given in the introduction.

5.2. Buildings over $\mathbf{F}_q$

Let k be a field and E a set. We define a structure of a k-projective line (resp. k-affine line; resp. k-projective plane) on E as a set of bijections of $\mathbf{P}_1(k)$ (resp. k; resp. $\mathbf{P}_2(k)$) onto E of the form $\tau \cdot PGL_2(k)$ (resp. $\tau \cdot Aff_1(k)$; resp. $\tau \cdot PGL_3(k)$), where τ is one such bijection. Here, $Aff_1(k)$ denotes the group of affine transformations $t \mapsto at + b$, $a \neq 0$, of k. We shall need the following observations:

(a) The dual of a k-projective plane is also, in a canonical way, a k-projective plane (k is commutative !).

(b) The lines of a k-projective plane have natural structures of k-projective lines.

(c) In a triangle building, the lines of the two projective planes attached to a vertex (namely the planes, duals of each other, whose flag complexes are the star of the vertex in question) are the stars of the 1-simplices containing that vertex.

Now, we define a *triangle building over* k as a triangle building in which the planes attached to all vertices are endowed with structures of k-projective planes and the stars of 1-simplices are endowed with structures of k-projective lines, all those structures being compatible with each other in the obvious sense (made clear by (a), (b), (c)), so that any single one of them determines all the others. For example, the affine building of the group SL_3 of a field with a discrete valuation is a triangle building over the residue field.

The spheres of radius 2 in triangle buildings over k are systems $\mathcal{S} = (\omega; P_1, P_2; C; (Q_x))$ as above in which, furthermore, P_1, P_2 are k-projective planes (with dual k-structures), the sets $C(x,y)$ are k-affine lines and all given k-structures are coherent, in the sense that they induce structures of k-projective planes on the planes Q_x. We are now interested in the number of isomorphism classes of such spheres for $k = \mathbf{F}_q$; so, let P_1, P_2 be two k-projective planes over $\mathbf{F}_q$, duals of each other, and for every flag $(a,b) \in P_1 \times P_2$, $a \perp b$, let there be given a k-affine line $C(a,b) = C(b,a)$. We consider the set $\mathcal{T}'$ of all spheres with k-structures, as above, made of the given data and varying plane structures Q_x.

In order to apply the same method as in 5.1 to the set $\mathcal{T}'$, one can start with the subgroup of Σ consisting of all elements preserving the given k-structures, but it is somewhat simpler to use a smaller group Σ', leading to the same result and defined as follows. For $x \in P_1$ ($i = 0$ or 1), let Σ'_x be the group of all permutations of $\bigcup\{C(x,y) \mid y \in P_{3-i}, x \perp y\}$ stabilizing each $C(x,y)$ and inducing on it an affine transformation, and let Σ' be the direct product of all Σ'_x, for $x \in P_1 \cup P_2$. Define the groups Σ'_0 and, for $\mathcal{S} \in \mathcal{T}'$, the groups $\Sigma'^{\mathcal{S}}_x$ and $\Sigma'^{\mathcal{S}}$ in the same way as Σ_0, $\Sigma^{\mathcal{S}}_x$, $\Sigma^{\mathcal{S}}$; in other, more explicit terms, set $\Sigma'_0 = \Sigma_0 \cap \Sigma'$, $\Sigma'^{\mathcal{S}}_x = \Sigma^{\mathcal{S}}_x \cap \Sigma'_x$, $\Sigma'^{\mathcal{S}} = \prod \Sigma'^{\mathcal{S}}_x = \Sigma^{\mathcal{S}} \cap \Sigma'$. The orders of the various groups we have introduced are:

$$\begin{aligned}
|\Sigma'_x| &= (q(q-1))^{q+1}\,, \\
|\Sigma'| &= (q(q-1))^{2(q+1)(q^2+q+1)}\,, \\
\left|\Sigma'^{\mathcal{S}}_x\right| &= q^2(q-1), \\
\left|\Sigma'^{\mathcal{S}}\right| &= \left(q^2(q-1)\right)^{2(q^2+q+1)}\,, \\
|\Sigma'_0| &= (q(q-1))^{(q+1)(q^2+q+1)}\,.
\end{aligned}$$

As in 5.1, the group Σ' operates transitively on $\mathcal{T}'$, one has a Proposition 2′, entirely similar to Proposition 2, and two Corollaries, 4′ and 5′, of which we state only the last one, as a final result:

Corollary 5′: *The number of isomorphism classes of spheres of radius* 2 *in triangle buildings over* $\mathbf{F}_q$ *is* $\geq$

$$\begin{aligned}
&|\Sigma'| \cdot \left|\Sigma'^{\mathcal{S}}\right|^{-1} \cdot |\Sigma'_0|^{-1} \cdot |PGL_2(\mathbf{F}_q)|^{-1} \\
&\qquad = q^{(q-3)(q^2+q+1)-3}(q-1)^{q^3-3}\left(q^2+q+1\right)^{-1}(q+1)^{-1}.
\end{aligned}$$

Remarks:

(a) One may be interested in triangle buildings *admitting* a structure of building over $\mathbf{F}_q$ without, however, including such a structure in the data one wants to enumerate. The effect of this weakening is to divide the expression in Corollary 5′ by n.

(b) If, in the first definitions of 5.2, one replaces the groups $PGL_1(k)$, $Aff_1(k)$ and $PGL_2(k)$ by their semi-direct extensions by $Aut\ k$, one defines objects that can be called semi-k-projective or -affine lines and semi-k-projective planes. Using them instead of the k-objects and copying the definition of a triangle building over k, one gets a class of buildings which, for lack of a better idea, I shall call here $\frac{1}{2}k$-triangle buildings. A triangle building has at most one $\frac{1}{2}k$-structure, so that the distinction appearing in Remark (a) between buildings *admitting* a k-structure and buildings *with* a k-structure has no equivalent for $\frac{1}{2}k$-buildings. An alternative definition of $\frac{1}{2}k$-triangle buildings is provided by the following criterion: *a triangle building has a* $\frac{1}{2}k$*-structure if and only if the stars of its vertices are flag complexes of* k*-projective planes and if, for every* 1*-simplex* (a,b)*, there exists an isomorphism of* Star a *onto* Star b *fixing (elementwise)* Star $\{a,b\}$.

We leave as an exercise the determination of a lower bound of the number of isomorphism classes of spheres of radius 2 in $\frac{1}{2}k$-triangle buildings by the same method as in 5.1 and 5.2.

References:

[1] P. Köhler, T. Meixner and M. Wester, The 2-adic affine building of type $\tilde{A}_2$ and its finite projections, *J. Combin. Theory Ser.* A 38 (1985), 203-209.

[2] ____, ____, ____, The affine building of type $\tilde{A}_2$ over a local field of characteristic 2, *Arch. Math.* 42 (1984), 400-407.

[3] M. Ronan, Triangle geometries, *J. Combin. Theory Ser.* A 37 (1984), 294-319.

[4] ____, A construction of buildings with no rank 3 residues of spherical type, *in* Buildings and the geometry of diagrams, Como 1984, *Lecture Notes in Math.* 1181, Springer Verlag, 1986, 242-248.

[5] ____, *Lectures on buildings*, Perspectives in Mathematics, vol. 7, Academic Press, 1989.

[6] J. Tits, Résumé de cours, *Annuaire du Collège de France*, 85e année (1984-1985), 93-110.

[7] ____, Buildings and group amalgamations, *in* Groups – St. Andrews 1985, *London Math. Soc. Lecture Notes* 121, Cambridge Univ. Press. 1986, 110-127.

[8] ____, Sur le groupe des automorphismes de certains groupes de Coxeter, *Journal of Algebra* 113 (1988), 346-357.

A Census of Finite Generalized Quadrangles

Stanley E. Payne

Abstract

A construction is given for each known isomorphism class of finite generalized quadrangles.

I. Prolegomena

Generalized quadrangles, as a special case of generalized polygons, were introduced by J. Tits in 1959 [26]. There he gave what are considered to be the classical examples. As special cases of other types of geometries, these classical examples appeared elsewhere, most notably among the partial geometries of R. C. Bose [3]. The first nonclassical examples were also found by J. Tits in the mid-1960s and first appeared in P. Dembowski [5]. All these examples had parameters (s,t) of the form (q,q), (q,q^2), (q^2,q), (q^2,q^3), or (q^3,q), where q is an arbitrary prime power. Then in 1969 R. W. Ahrens and G. Szekeres [1] constructed examples with (s,t) equal to $(q-1,q+1)$. For q a power of 2, these examples were found independently by M. Hall, Jr. [9]. Shortly afterwards, S. E. Payne [17] gave an abstractly more general construction which was shown in [18] to produce a new family for q any power of 2. More recently, starting with a construction of W. M. Kantor [10], several new families with parameters (q^2,q) have been discovered, often with restrictions on the prime power q.

Although the point-line dual of a GQ with parameters (s,t) is a GQ with parameters (t,s), implying that one could assume that s is never larger than t, we often find it interesting to mention both a GQ and its point-line dual. In any case, each finite GQ known to us is isomorphic to at least one of the constructions given below or to its point-line dual.

II. Classical Examples

The classical GQ are, by work of F. Buekenhout and C. Lefèvre [4], precisely the ones which may be embedded in $PG(d,q)$, $d = 3$, 4, or 5. The brief descriptions below are given with more detail in the monograph [21].

1. Let Q be a nonsingular quadratic in $PG(d,q)$ whose subspaces of maximal dimension are lines. Then the points of $\mathbf{Q}$ together with the lines of $\mathbf{Q}$ form a GQ, denoted $\mathbf{Q}(d,q)$, with parameters (q,q^k), where $k = d-3$, $d = 3$, 4, or 5.

2. Let $\mathbf{H}$ be a nonsingular hermitian variety of $PG(d,q^2)$, $d=3$ or 4. Then the points of $\mathbf{H}$ together with the lines on $\mathbf{H}$ form a GQ, denoted $\mathbf{H}(d,q^2)$, with parameters (q^2,q) or (q^2,q^3) according as $d=3$ or $d=4$.
3. The points of $PG(3,q)$, together with the totally isotropic lines of a symplectic polarity, form a GQ denoted $W(q)$ and having parameters (q,q).

The point-line dual of $\mathbf{Q}(4,q)$ is isomorphic to $W(q)$. The point-line dual of $\mathbf{Q}(5,q)$ is isomorphic to $\mathbf{H}(3,q^2)$.

III. Nonclassical Examples of J. Tits

Let $\mathbf{O}$ be an oval or ovoid of $PG(d,q)$, according as $d=2$ or 3. Then there is a GQ denoted $T_d(\mathbf{O})$ and constructed as follows: Let $PG(d,q)=P$ be embedded as a hyperplane in $PG(d+1,q)=G$. Define points as (i) the points of $G\backslash P$, (ii) the hyperplanes X of G for which X meets $\mathbf{O}$ in a unique point, and (iii) one new symbol (∞). Lines are defined as (a) the lines of G which are not contained in P and which meet $\mathbf{O}$ in a (necessarily unique) point, and (b) the points of $\mathbf{O}$. Incidence is defined as follows: a point of type (i) is incident only with lines of type (a); here the incidence is that of G. A point of type (ii) is incident with all lines of type (a) contained in it and with the unique point of $\mathbf{O}$ contained in it. The point (∞) is incident with no lines of type (a) and with all lines of type (b).

$T_2(\mathbf{O})$ has parameters (q,q), and $T_3(\mathbf{O})$ has parameters (q,q^2). When $\mathbf{O}$ is a conic, $T_2(\mathbf{O})$ is isomorphic to $Q(4,q)$. When $\mathbf{O}$ is an elliptic quadric, $T_3(\mathbf{O})$ is isomorphic to $Q(5,q)$. When q is odd, by a celebrated result of B. Segre [22] any oval $\mathbf{O}$ must be a conic. And by an analogous result of A. Barlotti [2] any ovoid $\mathbf{O}$ must be an elliptic quadric. So for q odd, this construction gives nothing nonclassical. But for q even, the ovoid discovered by J. Tits (cf. [25]) in $PG(3,q)$ does provide a nonclassical $T_3(\mathbf{O})$. And there are many nonconical ovals in $PG(2,q)$, q even, so that there are many nonclassical $T_2(\mathbf{O})$. We will mention these again a little later.

IV. The GQ $AS(q)$ of Ahrens-Szekeres with q Odd

Let q be any odd prime power. The construction given here yields a GQ isomorphic to that given via coordinates in [1] (cf. 3.2.6 of [21] for a proof).

Let $W(q)$ be the classical GQ arising from a symplectic polarity. Let x be any fixed point of $W(q)$, and let P be the polar plane of x with respect to the polarity. Points of $AS(q)$ are the points of $PG(3,q)$ not in P; lines of $AS(q)$ are the lines of $W(q)$ that do not contain x as well as all the lines of $PG(3,q)$ through x not lying in P. Incidence is that inherited from $PG(3,q)$. The resulting structure is a GQ with parameters $(q-1,q+1)$.

V. Hyperovals and GQ

Let $\mathbf{O}$ denote a hyperoval of $P=PG(2,q)$, q even. So $\mathbf{O}$ is a set of

$q+2$ points of P with no three collinear. Embed P as a hyperplane of $G = PG(3,q)$.

1. $(q-1, q+1)$. A GQ $T_2^*(\mathbf{O})$ is constructed as follows: Take as points the points of G not in P. And take as lines those lines of G not in P and meeting P in a unique point of $\mathbf{O}$. Incidence is that of G.

   This construction is just that of Ahrens-Szekeres [1] and of Hall [9].
2. (q,q). Let $\mathbf{n}$ be any point of $\mathbf{O}$, so $\mathbf{O_n} = \mathbf{O}\backslash\{\mathbf{n}\}$ is an oval with nucleus $\mathbf{n}$. We give an alternate description of the construction of $T_2(\mathbf{O_n})$. Points are the points of $G\backslash P$ and the planes of G containing $\mathbf{n}$. Lines are the lines of G tangent to $\mathbf{O_n}$. Incidence is the natural one inherited from G. For example, as a point of $T_2(\mathbf{O_n})$ P is incident with the lines of P through the point $\mathbf{n}$.
3. $(q+1, q-1)$. Let $\mathbf{m}, \mathbf{n}$ be any two distinct points of $\mathbf{O}$. Then $\mathbf{O}(\mathbf{m},\mathbf{n}) = \mathbf{O}\backslash\{\mathbf{m},\mathbf{n}\}$ is a q-arc, i.e., a set of q points, no three of which are collinear. (Conversely, any q-arc may be extended to a $q+2$-arc, in a unique way if $q \geq 4$. See J. A. Thas [23].) A GQ $\mathbf{S}(\mathbf{O}(\mathbf{m},\mathbf{n})) = \mathbf{S}$ is constructed as follows.

   Points are of three types: the points of $G\backslash P$, the planes of G that contain $\mathbf{n}$ but not $\mathbf{m}$, and the planes of G that contain $\mathbf{m}$ but not $\mathbf{n}$. The lines of $\mathbf{S}$ are the lines of G not in P but meeting P in a point of $\mathbf{O}(\mathbf{m},\mathbf{n})$. Incidence is the one naturally induced by that in G.

VI. A Garden of GQ

Two of the families of GQ given above have parameters of the form (q^2, q), viz., $H(3,q^2)$ (the point-line dual of $Q(5,q)$) and the point-line dual of $T_3(\mathbf{O})$, where $\mathbf{O}$ is the ovoid of J. Tits. In the first case, q is an arbitrary prime power, and in the second it is an odd power of 2. In addition to the GQ already described above, there are eight other known families of GQ, all with parameters of this general form and all discovered during the 1980s. The stories of these discoveries blend in an interesting way, connecting the study of translation planes, spreads of $PG(3,q)$, flocks of a quadratic cone, and even generalized hexagons! In this survey, we simply give one general account of their construction.

Let $\mathcal{S} = (P, B, I)$ be a GQ of order (s,t) with pointset P, lineset B, and incidence relation I. Let G be a group of collineations of S leaving invariant each line through a fixed point p and acting regularly on the set $P - p^\perp$ of points not collinear with p. Then $|G| = s^2t$. Fix $y \in P - p^\perp$, and let $p \, I \, L_i \, I z_i \, I \, M_i \, I \, y$, $0 \leq i \leq t$. Put $A_i = \{g \in G : M_i^g = M_i\}$, $A_i^* = \{g \in G : z_i^g = z_i\}$. Then $\{e\} < A_i < A_i^* < G$, $|A_i| = s$, $|A_i^*| = st$. Moreover, the family $\mathcal{F} = \{A_i : 0 \leq i \leq t\}$ is a *4-gonal family for* G, i.e.,

K1. $A_iA_j \cap A_k = \{e\}$, for i,j,k distinct,

and

K2. $A_i^* \cap A_j = \{e\}$, for $i \neq j$.

W. M. Kantor [10] was the first to observe these properties and the following converse. Let G be a finite group with $|G| = s^2t$, $s > 1$, $t > 1$. Let $\mathcal{F}$ be a family of $1+t$ subgroups of G, each having order s. For each $A \in \mathcal{F}$, let A^* be a subgroup of G containing A with $|A^*| = st$. Define as follows a point-line incidence geometry $\mathcal{S} = (P, B, I)$ with pointset P, lineset B, and incidence relation I, where:

Points are of three kinds:

(i) elements of G

(ii) right cosets A^*g $(A \in \mathcal{F},\ g \in G)$

(iii) one symbol (∞)

Lines are of two kinds:

(a) right cosets Ag $(A \in \mathcal{F},\ g \in G)$

(b) symbols $[A]$ $(A \in \mathcal{F})$

A point g of type (i) is incident with the line Ag for each $A \in \mathcal{F}$. A point A^*g of type (ii) is incident with $[A]$ and with each line Ahg contained in A^*g $(A \in \mathcal{F},\ g \in G,\ h \in A^*)$. The point (∞) is incident with each line $[A]$ of type (b). Then $\mathcal{S}$ is a GQ of order (s,t) if and only if $(G, \mathcal{F})$ satisfies K1 and K2, i.e., $\mathcal{F}$ is a 4-gonal family for G.

All the GQ to be described in this section arise from 4-gonal families in the manner just given.

Let $F = GF(q)$, q any prime power. Let $f : F^2 \times F^2 \to F$ be a fixed symmetric, nonsingular biadditive map. Put $G = \{(\alpha, c, \beta) : \alpha, \beta \in F^2,\ c \in F\}$. Define a binary operation on G by: $(\alpha, c, \beta) \cdot (\alpha', c', \beta') = (\alpha + \alpha', c + c' + f(\beta, \alpha'), \beta + \beta')$. This makes G into a group whose center is $C = \{(0, c, 0) \in G : c \in F\}$. Suppose that for each $t \in F$ there is an additive map $\delta_t : F^2 \to F^2$ and a map $g_t : F^2 \to F$ for which

$$g_t(\alpha + \beta) - g_t(\alpha) - g_t(\beta) = f(\alpha^{\delta_t}, \beta) = f(\beta^{\delta_t}, \alpha),$$

for all $\alpha, \beta \in F^2$, $t \in F$. With such a setup, we can define a family of subgroups of G by:

$$A(t) = \{(\alpha, g_t(\alpha), \alpha^{\delta_t}) : \alpha \in F\}, \quad t \in F,$$

and

$$A(\infty) = \{(0, 0, \beta) \in G : \beta \in F^2\}.$$

Then put $\mathcal{F} = \{A(t) : t \in F \cup \{\infty\}\}$. And for each $A \in \mathcal{F}$, put $A^* = AC$. So

$$A^*(t) = \{(\alpha, c, \alpha^{\delta_t}) : \alpha \in F^2\}, \quad t \in F,$$

and

$$A^*(\infty) = \{(0, c, \beta) \in G : \beta \in F^2\}.$$

Necessary and sufficient conditions on g_t were worked out in [15] (or see Section 10.4 of [21]) for $\mathcal{F}$ to be a 4-gonal family.

VI.1. $\mathcal{F}$ is a 4-gonal family for G if and only if

(i) $\delta(t,u) : \alpha \to \alpha^{\delta_t} - \alpha^{\delta_u}$ is nonsingular for $t \neq u$, so $\delta^{-1}(t,u)$ is well defined.

(ii) $g_t(\alpha) = g_u(\alpha)$, $t \neq u$, implies $\alpha = 0$.

(iii) If t, u, and v are distinct, then $\gamma = 0$ is the only solution to

$$g_t(\gamma^{\delta^{-1}(t,v)}) - g_v(\gamma^{\delta^{-1}(t,v)}) + g_v(-\gamma^{\delta^{-1}(v,u)}) - g_u(-\gamma^{\delta^{-1}(v,u)}) = 0.$$

Let $\mathcal{C} = \{A_t : t \in F\}$ be a set of q distinct 2×2 matrices over F. Then $\mathcal{C}$ is called a *q-clan* provided $A_t - A_u$ is anisotropic whenever $t \neq u$, i.e., $\alpha(A_t - A_u)\alpha^T = 0$ has only the trivial solution $\alpha = (0,0)$. For $A_t \in \mathcal{C}$, put $K_t = A_t + A_t^T$, and then define $g_t(\alpha) = \alpha A_t \alpha^T$ and $\alpha^{\delta_t} = \alpha K_t$ for $\alpha \in F^2$. With G, $A(t)$, $A^*(t)$, $\mathcal{F}$ as above, we have the following:

VI.2. (Combined results from W. M. Kantor [12] and S. E. Payne [15, 16].) $\mathcal{F}$ is a 4-gonal family for G if and only if $\mathcal{C}$ is a q-clan.

All but one of the known 4-gonal families arising from the general setup of VI.1 actually arise from q-clans as in VI.2, including the classical examples and those found by W. M. Kantor. To give these, it is sufficient to give A_t for each $t \in F$.

Example 1. (Classical; J. Tits [5]; cf. [21] for a proof that these are classical.)

Let $x^2 + bx + c$ be a fixed irreducible polynomial over F. Then put $A_t = \begin{pmatrix} t & bt \\ 0 & ct \end{pmatrix}$, $t \in F$.

Example 2. (W. M. Kantor [10]; associated with $G_2(q)$.)

Let $q \equiv 2 \pmod 3$. Then put $A_t = \begin{pmatrix} t & 3t^2 \\ 0 & 3t^3 \end{pmatrix}$, $t \in F$.

Example 3. (W. M. Kantor [12].)

Let q be odd, $\sigma \in \mathrm{Aut}(F)$, m a nonsquare of F, and put $A_t = \begin{pmatrix} t & 0 \\ 0 & -mt^\sigma \end{pmatrix}$, $t \in F$. These examples were studied further in [14]; also see VII.3 of [13].

Example 4. (W. M. Kantor [12] for q odd; S. E. Payne [16] for q even.)

Let $q \equiv 2 \pmod 5$. Put $A_t = \begin{pmatrix} t & 5t^3 \\ 0 & 5t^5 \end{pmatrix}$. This example has received little attention so far in case q odd. But for $q = 2^e$, e odd, an interesting byproduct was a previously unknown hyperoval in $PG(2,q)$.

Example 5. (J. A. Thas [24].)

Let $F = GF(q)$, $E = GF(q^2)$, q odd. Let ζ be a primitive element of E, so $w = \zeta^{q+1}$ is a primitive element of F and hence a nonsquare in F. Put $i = \zeta^{(q+1)/2}$, so $i^2 = w$, $i^q = -i$. Put $z = \zeta^{q-1} = a + bi$. Then $\langle z \rangle$ has order $q+1$ as a subgroup of E. For each k modulo $q+1$, put

$$a_k = \frac{(z^{k+1} - z^{-(k+1)}) - (z^k - z^{-k})}{z - z^{-1}} = \frac{z^{k+1} + z^{-k}}{z+1},$$

$$b_k = \frac{i[(z^{k+1} + z^{-(k+1)}) - (z^k + z^{-k})]}{z - z^{-1}} = \frac{i(z^{k+1} - z^{-k})}{z+1}.$$

When $t^2 - 2(1+a)^{-1}$ is a square in F, put $A_t = \begin{pmatrix} t & 0 \\ 0 & -wt \end{pmatrix}$. The remaining matrices in the q-clan are those of the form $\begin{pmatrix} -a_{2j} & b_{2j} \\ b_{2j} & -wa_{2j} \end{pmatrix}$, $0 \le 2j \le q-1$. These examples were found by J. A. Thas using flocks of a quadratic cone discovered by J. C. Fisher (cf. [6]). They were described geometrically in [24] and studied further in [19] and [20]. Also see [11].

Example 6. (H. Gevaert and N. L. Johnson [8].)

Let $q = 5^e$, k a nonsquare of F, and put $A_t = \begin{pmatrix} t & t^2 \\ 0 & k^{-1}t + 2t^3 + kt^5 \end{pmatrix}$, $t \in F$. These examples are derived from Kantor's "likeable" planes (cf. [8] for an explanation of this) using the connections discovered by J. A. Thas [24].

Example 7. (H. Gevaert and N. L. Johnson [8].)

Let $q = 3^r$, n a nonsquare of F, and put $A_t = \begin{pmatrix} t & t^3 \\ 0 & -(nt + n^{-1}t^9) \end{pmatrix}$, $t \in F$. These examples are derived from some semifield planes of M. Ganley [7]. See Section VII, Example 7, of [13] for a demonstration that the above description yields precisely the family of examples with $q = 3^r$ given in [8].

Example 8. (S. E. Payne [13].)

Let $q = 3^r$, n a nonsquare of F. Then for $t \in F$, $\gamma \in F^2$, put

$$g_t(\gamma) = \gamma \begin{pmatrix} t & 0 \\ 0 & -nt \end{pmatrix} \gamma^T + \left[\gamma \begin{pmatrix} 0 & t \\ 0 & 0 \end{pmatrix} \gamma^T\right]^{\frac{1}{3}} + \left[\gamma \begin{pmatrix} 0 & 0 \\ 0 & -n^{-1}t \end{pmatrix} \gamma^T\right]^{\frac{1}{9}}.$$

These examples, described via VI.1, etc., also ultimately derive from the semifield planes of M. Ganley [7], but more than just the ideas of [24] is required (cf. [13]). They are the only examples known to arise via VI.1 which are not associated with q-clans via VI.2.

Example 9. (J. Tits (cf. Dembowski [5]).)

For completeness, we list here also the point-line dual of $T_3(\mathbf{O})$, where $\mathbf{O}$ is the ovoid discovered by J. Tits [25] in $PG(3,q)$, $q = 2^e$, e odd.

References:

1. R. W. Ahrens and G. Szekeres, On a combinatorial generalization of 27 lines associated with a cubic surface, *Jour. Austral. Math. Soc.* 10 (1969), 485–492.

2. A. Barlotti, Un'estensione del teorema di Segre-Kustaanheimo, *Boll. Un. Mat. Ital.* 10(3) (1955), 498–506.

3. R. C. Bose, Strongly regular graphs, partial geometries and partially balanced designs, *Pacific J. Math.* 13 (1963), 389–419.

4. F. Buekenhout and C. Lefèvre, Generalized quadrangles in projective spaces, *Arch. Math.* 25 (1974), 540–552.

5. P. Dembowski, *Finite Geometries*, Springer Verlag, 1968.

6. J. C. Fisher and J. A. Thas, Flocks in $PG(3,q)$, *Math. Zeit.* 169 (1979), 1–11.

7. M. J. Ganley, Central weak nucleus semifields, *Europ. J. Combin.* 2 (1981), 339–347.

8. H. Gevaert and N. L. Johnson, Flocks of quadratic cones, generalized quadrangles, and translation planes, *Geom. Ded.*, 27, no. 3(1988), 301 - 317.

9. M. Hall, Jr., Affine generalized quadrilaterals, *Studies in Pure Math.* (ed. L. Mirsky), Academic Press (1971), 113–116.

10. W. M. Kantor, Generalized quadrangles associated with $G_2(q)$, *Jour. Comb. Theory* 29(A) (1980), 212–219.

11. W. M. Kantor, Generalized quadrangles and translation planes, *Algebras, Groups and Geometries* 3 (1985), 313–322.

12. W. M. Kantor, Some generalized quadrangles with parameters (q^2,q), *Math. Zeit.* 192 (1986), 45–50.

13. S. E. Payne, An essay on skew translation generalized quadrangles, *Geom. Ded.*, to appear.

14. S. E. Payne, A garden of generalized quadrangles, *Algebras, Groups and Geometries* 3 (1985), 323–354.

15. S. E. Payne, Generalized quadrangles as group coset geometries, *Congressus Numerantium* 29 (1980), 717–734.

16. S. E. Payne, A new infinite family of generalized quadrangles, *Congressus Numerantium* 49 (1985), 115–128.

17. S. E. Payne, Nonisomorphic generalized quadrangles, *Jour. Alg.* 18 (1971), 201–212.

18. S. E. Payne, Quadrangles of order $(s-1, s+1)$, *Jour. Alg.* 22 (1972), 97–119.

19. S. E. Payne, Spreads, flocks and generalized quadrangles, *J. Geom.* 33(1988), 113 - 128.

20. S. E. Payne, The Thas-Fisher generalized quadrangles, *Annals of Discrete Math.* (ed. M. Marchi) 37 (1988), 357–366.

21. S. E. Payne and J. A. Thas, *Finite Generalized Quadrangles*, Pitman Pub. Co., London, 1984.

22. B. Segre, Ovals in a finite projective plane, *Canad. Jour. Math.* 7 (1955), 414–416.

23. J. A. Thas, Complete arcs and algebraic curves in $PG(2,q)$, *Jour. Alg.* 2 (1987), 451–464.

24. J. A. Thas, Generalized quadrangles and flocks of cones, *Europ. J. Comb.* 8 (1987), 441–452.

25. J. Tits, Ovoïdes et groupes de Suzuki, *Arch. Math.* 13 (1962), 187–198.

26. J. Tits, Sur la trialité et certains groupes qui s'en déduisent, *Inst. Hautes Etudes Sci. Publ. Math.* 2 (1959), 14–60.

Finite Geometries via Algebraic Affine Buildings

William M. Kantor*

The purpose of this note is (1) to give a vague description of algebraic affine buildings, (2) to give a brief description of arithmetic groups, and then (3) to combine these subjects in order to discuss finite quotients of algebraic affine buildings modulo suitable arithmetic groups.

1. Algebraic affine buildings

I'll start with an example:

Let V be $\mathbf{Q}^n$, with standard basis $v_1, \ldots, v_n$. Let p be a prime, and write $\mathcal{O} = \{\frac{a}{b} \mid a, b \in \mathbf{Z}, p \nmid b\}$; this is a subring of $\mathbf{Q}$ such that $\mathcal{O}/p\mathcal{O} \cong GF(p)$. Let $\mathbf{G}(\mathbf{Q})$ denote the group $SL(n, \mathbf{Q})$ of $n \times n$ matrices over $\mathbf{Q}$ of determinant 1. The subgroup

$$B = \begin{pmatrix} \mathcal{O} & \mathcal{O} & \ldots & \mathcal{O} \\ p\mathcal{O} & \mathcal{O} & \ldots & \mathcal{O} \\ \ldots & \ldots & \ldots & \ldots \\ p\mathcal{O} & \ldots & p\mathcal{O} & \mathcal{O} \end{pmatrix}$$

of $\mathbf{G}(\mathbf{Q})$ is called an *Iwahori subgroup*. Note that B mod p is a Borel subgroup of $\mathbf{G}(\mathcal{O} \bmod p) = \mathbf{G}(GF(p)) = SL(n, p)$. (This type of notation is intended to be thought of in terms of matrix groups, in which entries are taken from the indicated rings.) Using $\mathbf{G}(\mathbf{Q})$ and B, an affine building Δ is obtained as follows (cf. [10]):

Δ is a simplicial complex;

a simplex of Δ is a proper subgroup of $\mathbf{G}(\mathbf{Q})$ containing a conjugate of B;

X is a face of Y iff $X \geq Y$.

The maximal simplexes (chambers) are just the conjugates of B. There are n types of vertices, which are the conjugates of the subgroups of matrices blocked as follows:

$$\begin{matrix} r\{ \\ n-r\{ \end{matrix} \begin{pmatrix} \overbrace{\mathcal{O}}^{r} & \overbrace{\frac{1}{p}\mathcal{O}}^{n-r} \\ p\mathcal{O} & \mathcal{O} \end{pmatrix}$$

* Supported in part by NSF grant DMS 87-01794 and NSA Grant MDA 904-88-H-2040.

for $r = 1, \ldots, n$; for example, when $r = n$ this group is just $SL(n, \mathcal{O}) = \mathbf{G}(\mathcal{O})$. The above group is the stabilizer in $\mathbf{G}(\mathbf{Q})$ of the *$\mathcal{O}$-lattice*

$$L_r = \sum_{i \leq r} \mathcal{O}\frac{v_i}{p} + \sum_{i > r} \mathcal{O}v_i$$

and also of cL_r for all $c \in \mathbf{Q}^*$. (An *$\mathcal{O}$-lattice* is an $\mathcal{O}$-submodule of V generated by a basis.) Therefore, vertices correspond to *lattice classes* $[L] = \{cL \mid c \in \mathbf{Q}^*\}$; moreover, every $\mathcal{O}$-lattice is the image of a unique L_r under $\mathbf{G}(\mathbf{Q})$.

Note that the star of $[L_r]$ is just the usual (spherical) building for L_r/pL_r, i.e., for $\mathbf{G}(GF(p)) = SL(n, p)$. The corresponding diagram reflects this fact: it is the extended Dynkin diagram of type $\tilde{A}_n$, having n nodes. (Hide any node and observe that the diagram A_{n-1} is left: the diagram of the building of $SL(n,p)$.) There is also an obvious dihedral group of graph automorphisms, generated by the usual graph automorphism of the $SL(n,p)$ building together with an n-cycle (produced, for example, by the diagonal matrix $\mathrm{diag}(1, 1, \ldots, 1, p)$).

All of the above goes through with the field $\mathbf{Q}_p$ of p-adic numbers in place of $\mathbf{Q}$, using the ring $\mathbf{Z}_p$ of p-adic integers in place of $\mathcal{O}$. *This produces an isomorphic building.*

This is an example of an *algebraic affine building*. The general case is as follows. (For a much more precise description, see [10].)

K	field with a complete, discrete valuation	(such as $\mathbf{Q}_p$)
$\mathcal{O}$	corresponding valuation ring	($\mathbf{Z}_p$ in the case of $\mathbf{Q}_p$)
π	uniformizer	(p in the case of $\mathbf{Q}_p$)
$k = \mathcal{O}/\pi\mathcal{O}$	residue field	($GF(p)$ in the case of $\mathbf{Q}_p$)
$\mathbf{G}(K)$	an absolutely simple, simply connected algebraic group over K of rank $\ell \geq 2$	
$\mathbf{G}(\mathcal{O})$	the corresponding group over $\mathcal{O}$: think in terms of matrix entries; this is assumed to be a *special* subgroup of $\mathbf{G}(K)$ (see below)	
$\mathbf{G}(\mathcal{O}) \to \mathbf{G}(\mathcal{O} \bmod \pi) = \mathbf{G}(k)$		
B	*Iwahori subgroup*, the preimage in $\mathbf{G}(K)$ of a Borel subgroup of of $\mathbf{G}(k)$	
Δ	*affine building* of $\mathbf{G}(K)$	

As above, Δ is a simplicial complex whose simplexes are the proper subgroups of $\mathbf{G}(K)$ containing conjugates of B, with X a face of Y iff $X \geq Y$. Chambers are the conjugates of B. The rank of Δ is $\ell + 1$: there are $\ell + 1$ different types of vertices. The requirement that $\mathbf{G}(\mathcal{O})$ is special amounts to the fact that the associated diagram is the extended Dynkin diagram corresponding to the Dynkin diagram of the finite group $\mathbf{G}(k)$; the group $\mathbf{G}(K)$ always contains such subgroups [10]. In the example given above, all the vertex stabilizers are special.

A second example may help in wading through these definitions. Let f be a quadratic form on a vector space V over $\mathbf{Q}_p$, of Witt index $\ell \geq 2$. The corresponding algebraic group $\mathbf{G}(K)$ is the commutator subgroup of the orthogonal group of f. (Actually, in order to conform with the above definitions I should take $\mathbf{G}(K)$ to be the associated spin group. However, the center of that spin group acts trivially on Δ, and hence can be ignored for our purposes.) Here, vertices correspond to classes $[L]$ of *suitable* $\mathbf{Z}_p$-lattices L in V. (In unpublished work, Rehmann and Scharlau have obtained necessary and sufficient conditions in order that a lattice L produces a vertex – not just for this orthogonal situation, but for all the classical groups.)

More concretely, consider the quadratic form $f = \sum_1^4 x_i x_{i+4}$ on $\mathbf{Q}_p^8$, with orthogonal group $\Omega^+(8, \mathbf{Q}_p)$. Let $e_1, e_2, e_3, e_4, f_1, f_2, f_3, f_4$ be the standard basis, so that with respect to the underlying bilinear form $(e_i, e_j) = 0 = (f_i, f_j)$ and $(e_i, f_j) = \delta_{ij}$. The vertices of Δ are the lattice classes containing the lattices

$$\begin{aligned}
L_1 &= \langle e_1, e_2, e_3, e_4, f_1, f_2, f_3, f_4\rangle_{\mathbf{Z}_p}\\
L_2 &= \langle \frac{e_1}{p}, e_2, e_3, e_4, pf_1, f_2, f_3, f_4\rangle_{\mathbf{Z}_p}\\
L_3 &= \langle \frac{e_1}{p}, \frac{e_2}{p}, e_3, e_4, f_1, f_2, f_3, f_4\rangle_{\mathbf{Z}_p}\\
L_4 &= \langle \frac{e_1}{p}, \frac{e_2}{p}, \frac{e_3}{p}, \frac{e_4}{p}, f_1, f_2, f_3, f_4\rangle_{\mathbf{Z}_p}\\
L_5 &= \langle \frac{e_1}{p}, \frac{e_2}{p}, \frac{e_3}{p}, \frac{f_4}{p}, f_1, f_2, f_3, e_4\rangle_{\mathbf{Z}_p}
\end{aligned}$$

The diagram is $\tilde{D}_4$, extended D_4:  with central node arising from L_3; the special vertices are those corresponding to the other nodes.

2. Arithmetic groups

Let $\mathbf{G}(K)$ be as above. A subgroup G of $\mathbf{G}(K)$ is *discrete* if G_x is finite for each vertex x of Δ; and G is *cocompact* if it has a finite number of chamber-orbits (or, equivalently, a finite number of orbits on some type of vertices). These definitions are motivated by the topology on Δ inherited from that of the field K.

More notation is needed in order to describe the basic example of a discrete cocompact subgroup of $\mathbf{G}(K)$; in parentheses is the special case of the rational field itself. (Note, however, that there is an additional possibility being ignored here, in which the ground field F has nonzero characteristic.)

F finite extension of $\mathbf{Q}$
v place (valuation) (p or ∞ in the case of $\mathbf{Q}$)
$K = F_v$ completion of F at v ($\mathbf{Q}_p$ or $\mathbf{R}$ in the case of $\mathbf{Q}$)
$\mathcal{O}_v$ corresponding valuation ring if v is finite ($\mathbf{Z}_p$ in the case of $\mathbf{Q}$)
$\mathcal{O}^v = \{a \in F \mid v'(a) \geq 0$ for all finite $v' \neq v\}$ ($\mathbf{Z}[\frac{1}{p}]$ in the case of $\mathbf{Q}$)
$\mathbf{G}(F)$ is embedded in $\mathbf{G}(F_v)$ in the obvious manner.
$G = \mathbf{G}(\mathcal{O}^v)$

Assume that $\mathbf{G}(F_{v'})$ is compact for all infinite places v'. *Fix* a finite place v. *Assume* that $\mathbf{G}(F_v) = \mathbf{G}(K)$ is noncompact and has rank $\ell \geq 2$. Then: *G is discrete and cocompact in* $\mathbf{G}(F_v)$.

For example, if f is a quadratic form over $\mathbf{Q}$ then $\mathbf{G}(\mathbf{Q})$ is the (derived subgroup of the) orthogonal group; the only infinite place is ∞, with $\mathbf{Q}_\infty = \mathbf{R}$. Then the compactness of $\mathbf{G}(\mathbf{Q}_\infty) = \mathbf{G}(\mathbf{R})$ simply asserts that f is a definite quadratic form (over $\mathbf{R}$). The group G is $\mathbf{G}(\mathbf{Z}[\frac{1}{p}])$ in case v corresponds to the prime p.

Returning to the general case, a subgroup H of $\mathbf{G}(F_v)$ is called an *arithmetic group* if $H \cap \mathbf{G}(\mathcal{O}^v)$ has finite index in both H and $\mathbf{G}(\mathcal{O}^v)$. A fundamental theorem of Margulis asserts that every discrete cocompact group H arises in essentially this manner (this requires the assumption that $\ell \geq 2$; cf. [14]). While there are even more general definitions of this sort, here it will only be necessary to consider $\mathbf{G}(\mathcal{O}^v)$.

Now assume that $\mathbf{G}(\mathcal{O}_v)$ is a special vertex of Δ as in the preceding section. Its conjugates consist of one of the vertex types of Δ. Consider the orbits of G on this type of vertices. The number of orbits is an integer h that is *independent* of the choice of v – subject to the condition that $\mathbf{G}(F_v)$ is noncompact and has rank ≥ 2 [5]. This integer is called the *class number* of $\mathbf{G}(F)$. This is a standard concept in the case of quadratic forms (where, however, it is called the spinor class number: the class number has a similar but slightly different meaning [3]).

Borel and Prasad [2] recently have shown that, given an integer h, there are only a finite number of pairs consisting of a field F and group defined over F for which the class number is h. In the special case of orthogonal groups this was proved in [9].

The case of greatest interest in finite group theory and finite geometry occurs when $h = 1$. Here, $\mathbf{G}(\mathcal{O}^v)$ is *transitive* on one of the vertex types of the building. Note the bizarre fact that, whereas the buildings arising from different places v are drastically different (the residue fields have different characteristics, and hence so do stars within the buildings), nevertheless transitivity for one v implies transitivity for all (suitable) v!

Example: Let $F = \mathbf{Q}$, $V = \mathbf{Q}_p^6$ and $f = \sum_1^6 x_i^2$. Then $h = 1$. This form is clearly positive definite over $\mathbf{R}$. The resulting building has diagram  if $p \equiv 1 \pmod 4$ (in which case all vertices are special) and  otherwise (in which case all vertices corresponding to end-nodes are special). In either case there are obvious graph automorphisms possible. In fact, the *full* group of transformations over $\mathbf{Z}[\frac{1}{p}]$ preserving f *projectively* is transitive on the set of *all* vertices if $p \equiv 1 \pmod 4$, and on the set of all end-node vertices for the remaining primes p. This is precisely the sort of example discussed in [6] in an *ad hoc* manner.

This situation suggests that one should look for some kinds of classification theorems. In a series of papers, using very detailed case arguments, G. L. Watson determined all definite quadratic forms over $\mathbf{Q}$ of class number 1 in dimension $n \geq 5$ (as indicated above, the notion of class number he used is not quite the same as what I have been using in the general case; but class number 1 for Watson implies class number 1 for us; the converse is probably true when $n \geq 5$, but this has yet to be proved). Watson showed that $n \leq 10$: see the references in [13] for the cases $n \geq 7$; his results for $n = 5$ and 6 remained unpublished at the time of his death in 1988.

In more recent work still in progress, R. Scharlau has considered the corresponding problem of determining those quadratic forms over algebraic number fields for which $h = 1$ (with h as defined above). He used [9] to show that $n \leq 14$ – strengthening the estimates $n \leq 34$ in [9] and $n \leq 18$ in [5] – and that the possibilities for the field F are severely limited if $n \geq 6$. This work did not use buildings, groups, or geometry. However, it should eventually have interesting applications to finite geometry, for reasons to be explained in the next section.

The above example with $F = \mathbf{Q}$ and $f = \sum_1^6 x_i^2$ is especially interesting when $p = 2$. Here, $\mathbf{G}(\mathbf{Z}_2)$ is the stabilizer in $\mathbf{G}(\mathbf{Q}_2)$ of the lattice $L = \mathbf{Z}_2^6$. Then the stabilizer in $G = \mathbf{G}(\mathbf{Z}[\frac{1}{2}])$ of L is just $\mathbf{G}(\mathbf{Z})$, which is a finite group of the form $2^5 A_6$. This acts on $L/2L$, which is a vector space over $GF(2)$. The form f mod 2 is actually linear, with kernel $H = \{\, (x_i) \in L \mid \sum x_i \equiv 0 \pmod 2 \,\}$. There is then a quadratic form $\frac{1}{2} f$ mod 2 induced on $H/2L$. This form is preserved by $\mathbf{G}(\mathbf{Z})$, the induced group being $A_6 \cong O(5,2)'$. Thus, $\mathbf{G}(\mathbf{Z})$ is chamber-transitive on the building produced on $H/2L$, which is in turn the star of the vertex $[L]$. In view of the transitivity of G on the vertices of the same type as $[L]$, it follows that G is chamber-transitive. This is one of the examples of chamber-transitive groups classified in the result in [7] – a result which, to a large extent, is subsumed by those contained in Meixner's paper for this conference.

3. Finite geometries

After all of the infinite groups and complexes appearing in the previous sections, it is now time to describe some implications for finite geometry.

If Δ and Δ' are two simplicial complexes then a map $\phi : \Delta \to \Delta'$ is called a *cover* if it is simplicial, onto, and for each vertex x the restriction $\phi_{St(x)} : St(x) \to St(x\phi)$ is an isomorphism. (There is a more general notion of 2-*cover* [11], but covers will suffice for the present purposes.)

The case of concern here is the one in which Δ is an affine building obtained as in the preceding sections. Then Δ has rank $\ell + 1 \geq 3$. Note that Δ' will be a complex that locally looks exactly like Δ: stars in Δ' will be isomorphic to stars in Δ. What do all the finite complexes Δ' look like? There is a discrete automorphism group A of Δ such that $\Delta' \cong \Delta/A$; that is, Δ' can be identified in a natural manner with the set of orbits of A on Δ. In particular, A is cocompact; it is the group of deck transformations of the cover [11].

However, not all discrete cocompact groups A produce a simplicial complex Δ/A: just consider the case in which A happens to be transitive on some type of vertices of Δ. This is one of the reasons for replacing simplicial complexes by chamber systems when discussing building-like geometries [11] (cf. [4]). Nevertheless, for purposes of finite geometry complexes are the appropriate things to aim for. Fortunately, there is no difficulty finding such complexes – and unfortunately there are simply too many of them:

If A is a discrete cocompact subgroup of $\mathbf{G}(F_v)$ then

1. *So is every subgroup of finite index;*
2. *A is residually finite: the intersection of all the normal subgroups of finite index is 1; and*
3. *There is a constant M (depending on $\mathbf{G}(F)$ and v) such that, if D is a subgroup of A of index at least M, then Δ/D is a simplicial complex and $\Delta \to \Delta/D$ is a cover.*

Here, 1 is easy, 2 is not difficult, and 3 is an observation of Tits [12]. The net effect of these facts is that, as already stated, there are too many possibilities for the simplicial complex Δ'. What is needed is some way to narrow the study of such complexes. Not surprisingly, transitivity properties provide at least one way to do this.

The following simple construction produces finite complexes Δ' with large induced groups. Start with a discrete group G transitive on some vertex type of Δ. Take any normal subgroup A of G such that G/A is finite and not too small. Then, by 2 and 3 above, $\Delta' = \Delta/A$ will be a simplicial complex on which G/A acts, and G/A will also be transitive on a vertex type.

Example: Let f be a definite quadratic form on a vector space over $\mathbf{Q}$, let $\mathbf{G}(\mathbf{Q})$ be as usual, and assume that $\mathbf{G}(\mathbf{Q})$ has class number 1. Let p be a prime, and assume that f has Witt index ≥ 2 over $\mathbf{Q}_p$ and that $\mathbf{G}(\mathbf{Z}_p)$ is the stabilizer of a vertex of Δ. Then $G = \mathbf{G}(\mathbf{Z}[\frac{1}{p}])$ is transitive on that vertex type. Now if $m > 1$ is an integer then let $A(m) = \{\, g \in G \mid g \equiv 1 \pmod{m} \,\}$. This is a normal subgroup of G, and $G/A(m)$ is usually just $\mathbf{G}(\mathbf{Z}/m\mathbf{Z})$. If m is sufficiently large then $\Delta/A(m)$ will be a complex that locally looks exactly like Δ: stars in $\Delta/A(m)$ will be isomorphic to stars in Δ.

Problem: *What properties does $\Delta' = \Delta/A$ have, either in the preceding situation, or in the more general situation in which A is a normal subgroup of a group G transitive on some vertex type, or in even more general contexts?*

Since there are evidently large numbers of finite geometries Δ' obtained even in the preceding example, it is not at all clear what sorts of properties one should look for. Perhaps there are some kinds of configuration theorems possible. At the moment, the only results are essentially asymptotic in nature. These are motivated by [8], which contains analogous results when Δ is the affine building (a tree) for $SL(2, \mathbf{Q}_p)$. *Assume* that $G = \mathbf{G}(\mathcal{O}^v)$ is transitive on the type of vertices one of which has stabilizer $\mathbf{G}(\mathcal{O}_v)$ in $\mathbf{G}(F_v)$.

1. *There is a constant C such that, if $A \lhd G$, G/A is finite and Δ/A is a complex, then the diameter of Δ/A is $\leq C \log_2(\#)$, where $\#$ is the number of vertices, or alternatively the number of chambers, of Δ (the constant C depends on which definition of $\#$ is used).*

Here, *diameter* refers to the usual diameter of either of two graphs: the 1-skeleton of Δ/A, or the chamber-graph of Δ/A. This result is asymptotically best possible, as a simple counting argument shows; and it is an easy consequence of results in [1]. However, even a bound on the constant C seems to be very difficult to compute.

2. *There is a computable constant C' such that, for all A as in* **1**, *the geometric girth of Δ/A is $\geq C' \log_2(\#)$.*

Here, the *geometric girth* is the length of the shortest circuit *not* homotopic to 0; homotopy refers either to the usual simplicial concept or to the one for chambers [11]. For example, in the simplicial case of the Example in Section 3, when $p = 2$ one can take $C' = \frac{1}{30}$.

At this point it should be clear that this subject is in its infancy, at least from the point of view of finite geometry. It is not yet clear what the most important questions are; there certainly are few tools available to study the geometries $\Delta' = \Delta/A$.

References:

1. N. Alon and V. D. Milman, λ_1, isoperimetric inequalities for graphs, and superconcentrators. J. Comb. Theory (B)38 (1985) 73-88.
2. A. Borel and G. Prasad, Finiteness theorems for discrete subgroups of bounded covolume in semi-simple groups (to appear).
3. J. W. S. Cassels, *Rational Quadratic Forms.* Academic Press, London 1978.

4. W. M. Kantor, Generalized polygons, SCABs and GABs, pp. 79-158 in *Buildings and the Geometry of Diagrams*, Proc. C.I.M.E. Session, Como 1984. Springer Lecture Notes No. 1181, 1986.

5. W. M. Kantor, Class number and transitive automorphism groups of algebraic affine buildings (in preparation).

6. W. M. Kantor, Reflections on concrete buildings, in *Geometries and Groups, Proc. Conf. Geometries, Finite and Algebraic* (Noordwijkerhout 1986). Geom. Ded. 25 (1988) 121-145.

7. W. M. Kantor, R. A. Liebler and J. Tits, On discrete chamber-transitive automorphism groups of affine buildings. Bull. AMS 16 (1987) 129-133.

8. A. Lubotzky, R. Phillips and P. Sarnak, Ramanujan graphs. Combinatorica 8 (1988) 261-277.

9. U. Rehmann, Klassenzahl einiger totaldefiniter klassischer Gruppen über Zahlkörpern, Dissertation, Göttingen 1971.

10. J. Tits, Reductive groups over local fields. Proc. Symp. Pure Math. 33 (1979), 29-69.

11. J. Tits, A local approach to buildings, pp. 519-547 in *The Geometric Vein. The Coxeter Festschrift*. Springer, New York-Heidelberg-Berlin, 1981.

12. J. Tits, Buildings and Buekenhout diagrams, pp. 309-320 in *Finite simple groups II*, Proc. Symp. Durham 1978. Academic Press, London 1980.

13. G. L. Watson, One-class genera of positive quadratic forms in seven variables. Proc. London Math. Soc. 48 (1984) 175-192.

14. R. J. Zimmer, Ergodic theory and semisimple groups. Birkhäuser, Boston 1984.

Groups Acting Transitively on Locally Finite Classical Tits Chamber Systems

Thomas Meixner

1. Introduction

Tits chamber systems admitting a transitive group of automorphisms have seen considerable interest in the last six or seven years. As a general reference I want to mention the surveys (lectures held at the Como conference in 1984) by Kantor ([6]) and Timmesfeld ([28]). Some notions have been developed, some methods such as the "amalgam method" have proved to be very powerful instruments since then; different notions used, however, cause people to use different sets of hypotheses, and sometimes the interested reader might have trouble to see how the numerous papers written fit together and whether the results achieved give a *complete* classification of the objects one is interested in.

The purpose of this survey is to state the classification theorem that has been achieved since Como, to give the list of "types" allowed by the theorem together with explicit examples found by many authors, finally describe the systems of subgroups used in the literature with their various relationships. Thereby, I will have to repeat parts of Timmesfeld's survey, but will do so in order to be as clear as possible.

Let us talk about definitions and notation. The objects we are interested in are connected locally finite classical Tits chamber systems C of finite rank that admit a transitive group of automorphisms G with the property that the chamber stabilizers in G are finite. For the sake of easy reading I will use an abbreviation for these hypotheses on the chamber system C, introducing the term T-SCAB for the objects under consideration. Of course, this is not meant to be another notion in addition to the ones existing in this area, but serves only the purpose of avoiding too voluminous hypotheses in the theorems.

Since many of the papers I have to recall are dealing with (strong) parabolic systems, and since a large portion of the (group-theoretic) proofs are naturally given in terms of parabolic systems, I have to define these systems of subgroups and describe their relationship to T-SCABs. I decided not to talk about Tits geometries and the corresponding group geometries that occur in the literature; theorems on T-SCABs always imply the analogous theorems on Tits geometries having the necessary finiteness and transitivity conditions. And in addition to that, with a little care, the theorems on Tits geometries that are important in our situation and appear in the literature can be reformulated to give theorems on T-SCABs, at least on non-tight ones.

(1.1) Definition. Let $C = (C, (P_i), i \in I)$ be a connected chamber system over the finite index set I. Assume C is a locally finite classical Tits chamber system, i.e. for all pairs (i,j) with $i \neq j$ the set of $\{i,j\}$-residues of C consists of pairwise isomorphic finite classical generalized $m(i,j)$-gons (just generalized digons for $m(i,j) = 2, M = (m(i,j))$ the corresponding Coxeter matrix), and assume some subgroup G of the automorphism group of C acts transitively on C and has the property that the stabilizer in G of a chamber in C is finite. Then $C = C(G)$ is called a *T-SCAB*.

The Coxeter matrix M is usually given as a diagram Δ, with the usual conventions, i.e. in the diagram Δ on the set of nodes I we draw a single (resp. double, triple, quadruple; resp. no) bond between the nodes i and j, if $m(i,j) = 3$ (resp. 4,6,8; resp. 2); forgetting about the "strength" of the bonds we get the graph of the diagram. Referring to the diagram or its graph without explicit distinction allows us to talk about connectedness of the diagram, or to make statements like: the diagram "is" a complete (bipartite) graph with only bonds of strength m. If all rank 2 residues of C have the same characteristic p, we say C has characteristic p. If the residues are defined over the same field, we will usually give this field of definition as well.

A T-SCAB $C = C(G)$ of rank n is, given any chamber c of C, of course, uniquely determined by the group G, the stabilizer B of c in G, and the (setwise) stabilizers $X_1, \ldots, X_n$ in G of the rank 1 residues of c. Then C is canonically isomorphic to the coset chamber system $C(G; B; X_1, \ldots, X_n)$. Therefore we may always give T-SCABs in the latter way. The groups B resp. X_i will be called *Borel group* resp. *rank 1 parabolics* of the system $\{B; X_1, \ldots, X_n\}$ or of the group G, if no confusion can arise. And we call $\{B; X_1, \ldots, X_n\}$ a *system of parabolic subgroups* in G defining the T-SCAB C. A subgroup of G generated by k different rank 1 parabolics of the system is called *rank k parabolic* of the system; for these we use the following notation. If a system of subgroups $X_i, i \in I$, of some group G and some nonempty subset J of I is given, we denote by X_J the group $< X_j, j \in J >$. If the subset is specified by its elements, we usually leave away the set brackets. Hence rank 2 parabolics are denoted by $X_{i,j}$.

The aim of an investigation of T-SCABs should be a description of all T-SCABs C and all the corresponding automorphism groups G.

Given a T-SCAB $C = C(G)$ it has a universal 2-cover, which is again a T-SCAB C (more interesting: if the spherical rank 3 residues are covered by buildings, this will be the chamber system of a building ([34])) with the same diagram. And the group G "is lifted" to a transitive group G on that building, such that the system of parabolic subgroups stays "the same": G is isomorphic to the amalgamated sum of the system defined by the Borel group, the rank 1 parabolics and the rank 2 parabolics of X, morphisms being the embeddings in X (see [35]). Therefore, a procedure to get a classification like the one mentioned above, could be:

determine all the possible systems of parabolic subgroups for a T-SCAB as sets of (isomorphism types of) groups together with the corresponding embeddings (in other words determine all possible amalgams, i.e. all T-SCABs up to "local isomorphism"), afterwards try and determine all possible projections.

The first part of this program is discussed below. The second one is hardly manageable; even if one knows exactly the building C, one still requires deep results (see Kantor's lecture, these proceedings), to get statements about *all projections*.

What we really want to describe is the determination of all T-SCABs up to "parabolic isomorphism", that is the determination of the isomorphism types of the Borel groups, rank 1 and rank 2 parabolics of the systems defining a T-SCAB. This classification is now complete, if one combines results of several authors. The corresponding classification up to local isomorphism consists of determining all ways the systems of groups can be amalgamated. This additional work has not been done for all types.

Let us start with an example.

(1.2) Example ([16]): Let $G = A_7$ and B some Sylow 2-subgroup of G. Then G has exactly four subgroups $X_1, \ldots, X_4$ containing B which are isomorphic to Σ_4. For suitable labelling we get the following T-SCABs.

(a) $C(G; B; X_1, \ldots, X_4)$ has diagram 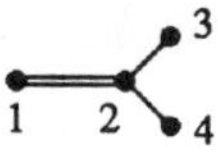

over $GF(2)$. It is called C_7^*.

(b) $C(G; B; X_1, X_2, X_i)$, $i = 3, 4$, have diagram 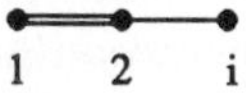

(C_3) over $GF(2)$. They are isomorphic, and are called C_7.

(c) $C(G; B; X_3, X_2, X_4)$ has diagram 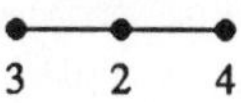

(A_3) over $GF(2)$. It is the chamber system of the $L_4(2)$-building.

(1.3) Example: Let G be a finite simple group of Lie-type of rank at least 2. Let B be some Borel subgroup of $G, X_1, \ldots, X_n$ the rank 1 parabolic subgroups of G containing B. Then the T-SCAB $C = C(G; B; X_1, \ldots, X_n)$

is the chamber system of the building adjoint to G and the diagram of the T-SCAB C is nothing else but the type of the building.

(1.3) gives of course most interesting examples, and T-SCABs are, in fact, just generalizations of the buildings of the finite simple Lie type groups. Therefore we want to have another look at this example.

Inside the group G, the system $X = \{B; X_1, \ldots, X_n\}$ of finite subgroups with B contained in the intersection of the X_i's has the following properties.

(1) $C(G; B; X_1, \ldots, X_n)$ *is a* T-SCAB.

(2) *There is a prime p such that for some Sylow p-subgroup S of B we have the following:*

- (i) $G = < X_1, \ldots, X_n >$,
- (ii) $S \in Syl_p(X_{i,j})$ *for all* i, j,
- (iii) *for all* i, $O^{p'}(X_i)/O_p(X_i)$ *is a central extension of a rank* 1 *Lie type group in characteristic* p,
- (iv) *for all* $i \neq j$, $O^{p'}(X_{i,j})/O_p(X_{i,j})$ *is a central extension of a rank* 2 *Lie-type group in characteristic* p.

(3) B *has a subgroup* H *(here it is a Cartan subgroup, which is a complement to* S *in* B*) such that for suitably chosen cosets* r_i *of* H *contained in* $X_i, i = 1, \ldots, n$, *and* $N := < r_i, i = 1, \ldots, n >$, *the quadruple* $(G, B, N, \{r_i, i = 1, \ldots, n\})$ *is a Tits system, hence the groups* B *resp.* X_i *are the group* B *resp. the rank* 1 *parabolics of a Tits system.*

Taking these three (sets of) properties as definitions, we get three concepts to look at systems $\{B; X_1, \ldots, X_n\}$ of finite groups generating a group G, that generalize the situation in the finite simple Lie type groups.

The first concept, T-SCABs, is the one we are talking about. The second kind of systems was given the name "strong parabolic systems" ([27]), and most of the papers recalled in the next section (mostly by Timmesfeld and Stroth) are concerned with this type (2) of systems and the slightly more general notion of "parabolic systems". A few words on the definition have to be added.

In the literature, the group B is dropped usually. The term "Lie type group" in (2) is used in the following way. Rank 1 Lie type groups in part (iii) of the definition are assumed to be finite simple Lie type groups of rank 1, or $L_2(2), L_2(3), U_3(2), Sz(2) = F_{20}, D_{10}$ or ${}^2G_2(3)$, which are not simple. The rank 2 Lie type groups in (iv) are the finite simple rank 2 Lie type groups plus the groups $Sp_4(2), G_2(2)$ and ${}^2F_4(2)$ together with their commutator subgroups, and direct products of rank 1 Lie type groups as just defined. So much for the definition of ***strong parabolic systems***.

The definition of a ***parabolic system*** coincides with (2) almost word by word; only in (iv), in the solvable cases some more possibilities are allowed to appear (certain subgroups of the corresponding rank 2 Lie type groups).

I do not have to say much about the concept of Tits systems (BN-pairs), which is considered in (3); this is the buildings' concept and here the classification theorem by Tits exists ([33]).

Let us add two more words on these three concepts. Systems of all three types are defined "locally", there is a (Coxeter) diagram attached to all of them, and (if these diagrams are connected) also a natural characteristic. And it is often possible to "view" systems of one type as systems of another one.

To be more precise, assume we have some system X of a given type in the group G and assume the corresponding diagram (type) is connected (irreducible). Then there are the following connections, recalled in large parts from [28].

(1.4) *Assume X has type* (3). We may assume that there is no nontrivial normal subgroup of G contained in B . Since B is finite, also the Weyl group of the Tits system must be finite ([27], (2.7)), hence the diagram is spherical and we may apply Tits's classification [33]. Then X is also of type (1) and (2) as already stated.

(1.5) *Assume X has type* (2), *so X is a* (*strong*) *parabolic system in characteristic p.* Then from $X = \{S; X_1, \ldots, X_n\}$ generating the group G we get a system $\{B; P_1, \ldots, P_n\}$ for a T-SCAB in some subgroup of G with the same diagram following ([28], [27],(2.6)). Put $Y_i = O^{p'}(X_i), B_i = N_Y(S)$ and $B =< B_i, i = 1, \ldots, n >, P_i = B.Y_i$. Then, the system $P = \{B; P_i, i = 1, \ldots, n\}$ gives a T-SCAB for the subgroup $G_0 =< Y_i, i = 1, \ldots, n >$ of G (after factoring out the largest normal subgroup of G_0 contained in B), and the system P is again a parabolic system for G_0. If moreover the parabolic system P is strong, and the rank 2 groups $P_{i,j}/O_p(P_{i,j})$ resp. the rank 1 groups $P_i/O_p(P_i)$ satisfy some conditions (*) resp. (**) given in [17], the system P is even a system of type (3) ([17]).

(1.6) *Assume X has type* (1). Then under some hypothesis on the diagram and excluding some explicitly given systems, there are subgroups B^* of B , X_i^* of X_i such that for $G^* =< X_i^*, i = 1, \ldots, n >$ the system $C(G^*; B^*; X_i^*, i = 1, \ldots, n)$ is a T-SCAB covering the T-SCAB defined by the system X in G, and $B^*; X_i^*, i = 1, \ldots, n$ is a parabolic system in G^*. ([27],Theorem 1, [28],(3.1)).

Thus, if for instance we want to show that a T-SCAB with a given non-spherical diagram cannot exist, we may try and apply (1.6),(1.5) and (1.4) in succession to get a contradiction. This procedure was introduced by Timmesfeld in the proof of his large order theorem ((2.2) below).

This attempt may, however, fail for several reasons. Firstly, there are the exceptions of (1.6), one has to take care of; then the parabolic systems one gets might not be strong; and finally Niles' conditions might not be satisfied in the rank 2 groups. In these cases, especially if some residues are defined over the field with two elements, quite different methods must be applied.

2. Classification theorems

In this section, we recall theorems on T-SCABs $C = C(G)$, which will yield a classification of T-SCABs up to parabolic isomorphism, i.e. giving the possible diagrams, characteristics and the rank 2 parabolics up to isomorphism. As a general hypothesis of this section we assume we are given a T-SCAB $C = C(G)$ defined by the system $X = \{B; X_1, \ldots, X_n\}$ in the group G. In view of [26],(4.4) and (4.5), we will assume that the diagram Δ of the T-SCAB C is connected, hence C has a well-defined characteristic p. Finally, we assume that the rank n of C is at least 3, referring to [21] for a classification of the rank 2 case.

The list of classification theorems for SCABs begins with the spherical case.

(2.1) Theorem. *Assume C is spherical. Then one of the following holds:*

(i) *Δ is of type C_3 and C is isomorphic to C_7, G is isomorphic to A_7.*

(ii) *C is the chamber system of a building of a finite simple group L of Lie type and G is contained in $Aut(L)$. Further G contains L or else C is the $A_3(2)$-building and G is isomorphic to A_7* (see (1.2)(*c*)).

A proof can be found in [30](3.1), see [28](2.11). The proof uses a theorem of Aschbacher ([1]) in the case some rank 3 residue of type C_3 is not a building, giving case (i). If all rank 3 residues of type C_3 are covered by, and hence are buildings, Tits' universal 2-cover theorem ([34]), the classification of buildings of spherical type ([33]) and the theorem by Seitz describing the flag-transitive subgroups of the Lie-type groups are the main ingredients leading to case (ii).

The idea, that non-spherical T-SCABs are a "small prime phenomenon", supported by the known examples, became a fact, when Timmesfeld proved his large-order-theorem.

(2.2) Theorem. *Assume all rank 1 residues of C contain at least 6 chambers. Then one of the following holds:*

(i) *C is a finite building and G is an extension of a group of Lie type by diagonal and field automorphisms.*

(ii) *Δ is a complete graph with only single bonds, all rank 1 residues are of size 9 (hence all rank 2 residues are projective planes over $GF(8)$) and for $i \neq j, X_{i,j}$ is a Frobenius group of order 73.9.*

(iii) $\Delta =$ 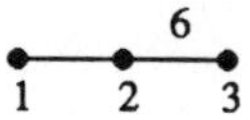

, $X_{2,3} = G_2(5), X_{1,2}$ *is a non-split extension of* $SL_3(5)$ *by an elementary abelian group of order* 5^3, *and* $X_{1,3} \cong 5^{1+4}Z$, *where* Z *is a central extension of* Σ_6 *by a group of order* 4.

(iv) $\Delta =$ 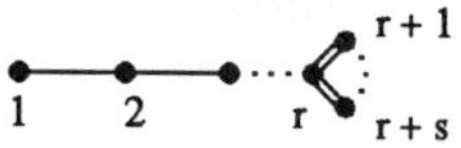

, *which we want to call* $Y(r,s), r \geq 1, s \geq 2$ *and the groups* $X_{1,2,\ldots,r,r+i}$ *are isomorphic to* $PSp_{2r+2}(7)$ *for* $1 \leq i \leq s$.

The proof of this theorem ([27],Theorem 1) consists ultimately in showing that in the remaining cases one has a strong parabolic system in G and by Niles' theorem ([17]) G must even have a BN-pair , thus can be obtained using Tits' classification. The proof is, in fact, the model for the discussion in section 1 above.

By the large order theorem, the only characteristics that have to be looked at further are 2 and 3. The most difficult case, which is also most important, since it is used inductively in all the higher rank situation, is the rank 3 case.

First, Timmesfeld handled the rank 3 case in characteristic 2 ([29],[31]; see [28],(3.4)) – later, the result was generalized to arbitrary characteristic in [32]. We give the characteristic 2 version here, since we want to treat characteristics 2 and 3 separately. To make the statements of the theorems to follow not too long, we will from now on not always give the exact structure of the rank 2 parabolics, but sometimes give only the parabolic type of the system by referring to a finite known example described in the literature.

(2.3) Theorem. *Let* C *be of rank* 3 *and characteristic* 2. *Then one of the following holds:*

(i) Δ *is spherical and* (C,G) *is as in* (2.1).

(ii) $\Delta =$  *and either* $X_{i,j}$ *is a Frobenius group of order* 21 *for all* $i \neq j$, *or a Frobenius group of order* 73.9 *for all* $i \neq j$.

(iii) $\Delta =$ 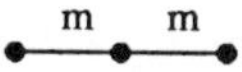

, $m = 4,6,8$, *and* $X_{1,2} \cong X_{2,3}$ *is an extension of a finite simple group of Lie type defined over* $GF(2)$ *by field automorphisms.*

(iv) $\Delta =$ 

and G *is parabolic isomorphic to* $G_2(3)$.

(v) $\Delta =$ 

and G *is parabolic isomorphic to* $G_2(3)$.

(vi) $\Delta =$  *and* G *is parabolic isomorphic to some subgroup of*

$PO_6^-(3)$ containing $P\Omega_6^-(3)$.

(vii) $\Delta =$ •═•═• *and G is parabolic isomorphic to the sporadic Suzuki group. (Labelling of the nodes in string diagrams $1, 2, 3$ from left to right.)*

The theorem is more or less a theorem on parabolics systems, the only exceptions occurring in (ii). The difference between the case (v) and the corresponding one of the cases in (iii) lies in the structure of the group $X_{1,3}$, which is 2-constrained in one case (v) and not 2-constrained in all the other cases of (iii).

The next theorem by Timmesfeld, which treats arbitrary rank in characteristic 2, but assumes that there are only single bonds in the diagram, was actually proved earlier than (2.3), it is the main theorem of [26], see ([28],(3.3)).

(2.4) Theorem. *Assume C has characteristic 2, and the diagram Δ contains only single bonds. Then one of the following holds:*

(i) *Δ is spherical, and (C, G) is as in (2.1).*

(ii) *Δ is a complete graph with only single bonds, and for $i \neq j, X_{i,j}$ is always a Frobenius group of order 21, or always a Frobenius group of order 73.9.*

(iii) *Δ is* ✕ *, and $X_{i,j,k,l} \cong \Omega_8^+(2)$ for all 4-element sets $\{i, j, k, l\}$ containing the middle node.*

(iv) *Δ is* □ *, and $X_{i,j,k} \cong A_7$ for all 3-element sets $\{i, j, k\}$.*

Knowing the rank 2 groups of the system (by definition), and also the rank 3 groups locally, and using Timmesfeld's classification if the diagram contains only single bonds – Stroth achieved the classification in characteristic 2 for arbitrary rank. (There are again papers that treat situations with one or the other hypothesis on the diagram, but I will restrict myself to the following general ones.)

(2.5) Theorem. *Assume the characteristic of C is 2 and there is a bond of strength at least 3 contained in the diagram. Then one of the following holds:*

(i) *Δ is a complete bipartite graph containing only bonds of the same strength $m \geq 3$, and all groups $X_{i,j}$ acting on a rank 2 residue which is not a generalized digon, are pairwise isomorphic of one of the following groups: $G_2(2), G_2(2)', {}^3D_4(2), {}^2F_4(2), {}^2F_4(2)'$.*

(ii) $\Delta =$ •—•$\overset{6}{—}$• *and G is parabolic isomorphic to $G_2(3)$.*

(iii) $\Delta =$ [diagram: nodes 1, 2, with bonds labelled 6 to nodes 3, …, r], $r \geq 4$, *and $X_{1,2,i}$ is parabolic isomorphic to $G_2(3)$ for all $i \in \{3, \ldots, r\}$.*

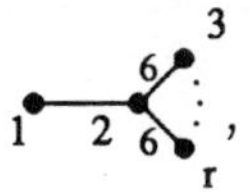

This theorem is the main theorem of [24]. I point out that due to the fact, that for the diagram •$\overset{6}{—}$•$\overset{6}{—}$• in (2.3) there are T-SCABs which are not parabolic isomorphic, the diagram does not completely determine the parabolic isomorphism type in (i) and (iii), but the parabolic isomorphism types can be derived easily.

Now, for the rest of the characteristic 2 discussion, one may assume that only bonds of strength 1 or 2 are involved in the diagram. However, the possible occurrence of C_7 as a rank 3 residue of type C_3 forces another subdivision.

The next theorem by Stroth ([23]) classifies the C_7-free case.

(2.6) Theorem. *Assume the characteristic of C is 2, and there are no bonds of strength at least 3 contained in the diagram. Assume that on rank 3 residues of type A_3 or C_3 there is always a full Lie-type group induced (i.e. they are all buildings and in case of the $A_3(2)$-building not just A_7, but $A_8 \cong L_4(2)$ is induced). Then one of the following holds:*

(i) *Δ is spherical and G, C are described in (2.1).*

(ii) *There is no double bond contained in Δ and (C,G) is as in (2.4)(iii) or (ii).*

(iii) *Δ is of rank 3 and (C,G) are described in (2.3).*

(iv) *Δ is of type $Y(s,r), r \geq 2, s \geq 1$, and $X_{1,.s,i}$ is isomorphic to $Sp_{2s+2}(2)$, $U_{2s+2}(2)$ or $U_{2s+1}(2)$.*

(v) *Δ is a complete bipartite graph containing only double bonds and rank 3 parabolics with connected diagram are parabolic isomorphic to one of the T-SCABs in (iii).*

(vi) *Δ is a tripartite graph with only double bonds on the node set $1\dot{\cup}I_1\dot{\cup}I_2$, such that $X_{1,i,j}$ is parabolic isomorphic to (2.3)(vii) for all $i \in I_1, j \in I_2$.*

I did not give all the information Stroth gives in case (v) to be not too long. If certain rank 2 parabolics occur, for instance, the complete bipartite graphs have to be stars.

If there is A_7 induced on some rank 3 residue of type A_3 or C_3, the classification is also complete. Since in the papers by Stroth and Heiss

([22],[4]) tight systems are excluded, resp. geometries are considered, I state the theorem with the hypothesis that the chamber systems are not tight. Recall that a rank n chamber system is called tight , if it equals one of its rank $n-1$ residues.

(2.7) Theorem. *Assume the characteristic of C is 2, and there are no bonds of strength at least 3 contained in the diagram. Assume that C is not tight, and assume on some rank 3 residue just A_7 is induced. Then one of the following holds:*

(i) *$C = C_7$ or the $L_4(2)$-building and $G = A_7$.*

(ii) *Δ is* [diagram: nodes 1, 2, 3, …, r] $, r \geq 4, X_{1,2,3} \cong 2^6 A_7, X_{1,2,i} \cong Sp_6(2), i \geq 4.$

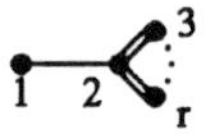

(iii) *Δ is* [diagram: nodes 1, 2, 3, 4] $, X_{1,2,3} \cong X_{1,2,4} \cong Sp_6(2), X_{3,2,4} \cong 2^6 A_7.$

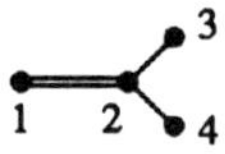

(iv) *Δ is* [diagram: nodes 1, 2, 3, 4, …, r] *, $r \geq 5$, and $X_{1,2,3,i}$ is parabolic isomorphic to some T-SCAB in* (iii).

(v) *Δ is* [diagrams] *and $X_{i,j,k} \cong A_7$ for all rank 3 parabolics of type A_3 or C_3.*

(vi) *Δ is* [diagram: nodes 1, 2, 3, 4] $, X_{1,2,3} \cong X_{1,4,3} \cong 2^4 A_7$ *and* $X_{2,1,4} \cong \widehat{2A_8}.$

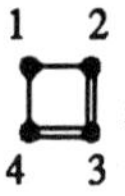

A proof of (2.7) in the case that the A_7 is induced on a residue of type A_3 is Theorem B of [4], while the proof of the Theorem in [22] applies in the case that A_7 is induced on some residue of type C_3.

The general version of (2.7) is very easily obtained; one has to determine the possibilities to introduce new parabolics in a given system to get new – higher rank, but tight – systems. The typical example is the system C_7*, which can be viewed as a tight extension of its rank 3 residues of types A_3 and C_3; for easy reference, before stating the corollary, I want to recall two tight rank 6 chamber systems.

Examples: (i) *Let G be the group $\Omega_6^-(3)$,* then G has subgroups $S \cong D_8, X_i \cong \Sigma_4, i = 1, \ldots, 6$, such that $C = C(G; S; X_1, \ldots, X_6)$ is a T-SCAB with diagram [diagram]. Label the nodes of the diagram $1, \ldots, 6$ clockwise (same convention in (ii) below), starting at the upper left node. All rank 3 parabolics of type C_3 are isomorphic to A_7, all rank 4 residues with 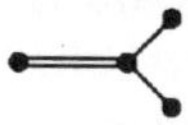 are isomorphic to C_7^*, in particular tight; there are, however, also rank n residues for $n = 4$ that are not tight T-SCABs: for $J = \{1,2,4,5\}$ resp. $\{1,2,3,4\}, \{1,2,3,6\}$ the corresponding diagrams for $C(X_J; S; X_i, i_J)$ are ,  resp. . For the rank 5 residues with $J=\{1,2,3,4,5\}$ resp. $\{1,2,3,4,6\}$, the diagrams are  and . We use the name O_J for the system on the set J for easy reference.

(ii) Let G be the group $U_3(5)$. Then G has subgroups $S \cong D_8, X_i \cong \Sigma_4, i = 1, \ldots, 6$, such that $C = C(G; S; X_1, \ldots, X_6)$ is a T-SCAB with diagram: 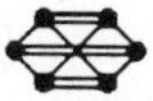 .

All rank 3 parabolics of type C_3 and A_3 are isomorphic to A_7, all rank 4 residues with diagram 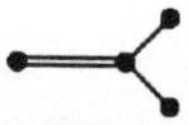 are isomorphic to C_7^*. Like in (i), we consider rank 4 resp. rank 5 residues to get T-SCABs with various diagrams. For $J = \{1,2,4,5\}$ resp. $\{1,2,5,6\}$ resp. $\{1,2,3,4,5\}$ we get the diagrams , ,  respectively. We call these T-SCABs U_J for easy reference.

The T-SCABs in (i) were described in [15], the T-SCAB U_{1245} was first described in [16], and the remaining T-SCABs in (ii) were constructed in [36]. Notice that the mentioned rank 4 T-SCABs appeared already in (2.7).

(2.8) Corollary. *Assume the characteristic of C is 2, and on some rank 3 residue just A_7 is induced. Then one of the following holds:*

(i) *(C, G) is as in (2.7).*

(ii) *C is C_7^* and $G = A_7$.*

(iii) *C is parabolic isomorphic to O_J for $J = \{1,2,3,4,5\}, \{1,2,3,4,6\}$ or $\{1,2,3,4,5,6\}$ in the example above.*

(iv) *C is parabolic isomorphic to U_J for $J = \{1,2,3,4,5\}$ or $\{1,2,3,4,5,6\}$ in the example above.*

The proof is an easy exercise.

The case that is left to be considered is the one of characteristic 3.

As mentioned above, the classification of parabolic systems in characteristic 3 and rank 3 is contained in [32], and so in the "parabolic-system-case" one knows pretty well what rank 3 parabolics look like. In characteristic 3, there are, however, some exceptional flag-transitive subgroups of rank 2 Lie-type groups ([21]); hence classifying all the T-SCABs of rank 3 in characteristic 3 needs a little extra work.

(2.9) Theorem. *Assume the characteristic of C is 3. Then one of the following holds:*

(i) *Δ is spherical, C is the chamber system of a building of type Δ and G is an extension of a finite simple Lie-type group by diagonal and field automorphisms.*

(ii) *Δ is a complete bipartite graph with only triple bonds. For $m(i,j) = 6$, the groups $X_{i,j}$ are pairwise isomorphic either to $G_2(3)$ or to ${}^3D_4(3)$ or $Aut({}^3D_4(3))$. In the second case Δ is a star.*

(iii) *Δ is of type $Y(s,r), r \geq 2, s \geq 1$, and $X_{1,..s,i}$ is isomorphic to one of $PSp_{2s+2}(3), \Omega_{2s+3}(3)$ or $U_{2s+2}(3)$ independent of $i > s$, (possibly extended by diagonal or field automorphisms).*

(iv) *Δ is a complete bipartite graph containing only double bonds, and if $m(i,j) = 4$, the group $X_{i,j}$ is isomorphic to $3.SO_5(3), SO_5(3)$ or $\Omega_5(3)$.*

(v) *Δ is a star containing only double bonds and if $m(i,j) = 4, X_{i,j}$ is isomorphic to one of the exceptional flag-transitive subgroups of $Aut(U_4(3))$ (see [21]).*

(vi) *Δ is a bipartite graph containing only double bonds, and if $m(i,j) = 4, X_{i,j}$ induces one of the exceptional flag-transitive subgroups of $Aut(\Omega_5(3))$ on its residue such that the kernel of this action is a $\{2,3\}$-group centralized by $X'_{i,j}$ (see [21]).*

The proof of (2.9), which uses some results from [32] for the rank 3 case, is given in [14]; a partial result was obtained before by Thiel ([25]). Instead of making general comments on it, I will sketch the proof in a typical situation.

Claim: Let $C = C(G)$ be a T-SCAB with characteristic 3, rank at least 3 defined by the system of subgroups $\{B; X_1, \dots, X_n\}$ of G; then Δ is not of type $\tilde{A}_{n-1}$.

Proof: Assume the contrary. A look at the diagram tells us that all rank 1 parabolics of the system are defined over the same field, and hence we may assume over $GF(3)$ by (2.4). Of course, to reach a contradiction we may replace the groups X_i according to (1.6) and (1.5), whose hypotheses are easily verified. Hence we may assume the X_i are in fact $\{2,3\}$-groups containing B with index 4, and B is 3-closed with Sylow 3-subgroup, say, S.

(i) *The rank is* 4.

Note that for i, j with $m(i,j) = 3$ the group $X_{i,j}$ induces $L_3(3)$ on its residue; and rank 1 parabolics in $L_3(3)$ induce $PGL_2(3)$ on their rank 1 residues, hence Niles condition (**) is satisfied in the rank 1 parabolics of the system . And it is easily checked that Niles condition (*) is also satisfied in $L_3(3)$. Hence we are done by the remarks in section 1, if we can verify (*) also in the rank 2 parabolics of type $A_1 \times A_1$. But we can do this, if it is possible to embed the corresponding rank 2 parabolics (residues) into rank 3 parabolics (residues) of type $A_1 \times A_2$, where the property is easily checked. This embedding, however, is always possible – except in the case of rank 4, hence the claim.

Put for convenience $G_i = X_{jkl}$ for $\{i,j,k,l\} = \{1,2,3,4\}$, and put $Z_i = \Omega_1(Z(O_3(G_i)))$. We will in the end get a contradiction playing the groups G_i against each other. We use, that $O^2(G_i/O_3(G_i))$ is isomorphic to $PSL_4(3)$ or $SL_4(3)$ by (2.1), and the faithful action of G on C, i.e. the fact that there is no nontrivial normal subgroup of G contained in B, in other words no nontrivial subgroup of B is invariant under two of the groups G_i.

(ii) *At least three of the Z_i are nontrivial modules for $O^2(G_i)$.*

Assume first that $O_3(G_i) = 1$ for some i. Then we know that S has 3^6 elements, and so $O_3(G_j) = 1$ for all j. But B contains exactly one elementary abelian subgroup of order 3^3 normal in X_1, as can be seen in G_2, hence the groups $O_3(X_{1,2})$ and $O_3(X_{4,1})$ must be equal, contradicting the structure of G_3.

Hence $M_i := O_3(G_i) \neq 1$ for all i, and clearly Z_i is an $(S)L_4(3)$-module for G_i.

Assume $Z(S)$ is not contained in M_1. Then G_1 is the product of B and the centralizer of M_1 in G_1, hence $Z_i \cap M_1$ is G_1-invariant for every i, and so trivial action of $O^2(G_i)$ on Z_i forces $Z_i \cap M_1 = 1$. This contradicts the structure of $G_i \cap G_1/M_1$. Now the claim holds.

Hence $Z(S)$ is contained in all M_i, and so if Z_i is a trivial module for $O^2(G_i)$ for two different values of i, we get $\Omega_1(Z(S))$ is normal in G, a contradiction. Therefore the claim holds in any case.

We may therefore assume that Z_1 and Z_3 are nontrivial $O^2(G_1)$- resp. $O^2(G_3)$-modules corresponding to the diagram (square with vertices 1 2 / 4 3) .

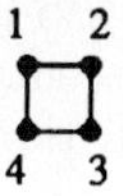

(iii) *Z_1 or Z_3 involves a natural $SL_4(3)$-module for G_1/M_1 resp. G_3/M_3.*

Since G is generated by G_1 and G_3, every element of G can be written in a shortest way as a product of elements of these two groups. Assume all conjugates of Z_1 are contained in $M_1 \cap M_3$. Then the normal closure of Z_1 in G is contained in S, a contradiction. The same argument shows that there are also conjugates of Z_3 not contained in M_1 or M_3. Now the minimum d of the lengths of expressions in elements of G_1 and G_3 of elements g in G with the property that there are i and j in $\{1,3\}$ such that Z_i is not contained in M_j^g, is well defined. And if we take a triple (i, j, g) where the minimum d is attained, we can easily see that Z_i is contained in G_j^g but not in M_j^g, and Z_j^g is contained in G_i but not M_i. But now one of Z_i and Z_j^g is a so called FF-module for the corresponding group G_i/M_i resp. G_j^g/M_j^g, which means that the group induced on the module contains an elementary abelian p-subgroup (p the characteristic of the module: hence $p = 3$ in this case) whose order is at least equal to the index of its centralizer in the module. If for instance $|Z_i : Z_i \cap M_j^g| \leq |Z_j^g : Z_j^g \cap M_i|, Z_i$ is a FF-module for G_i/M_i, and the elementary abelian 3-group needed is $Z_j^g M_i/M_i$. Now it is clear that Z_1 or Z_3 is also a FF-module for G_1/M_1 resp. G_3/M_3, since the defining property is unchanged by conjugation. And by the construction the "offending" elementary abelian 3-subgroup in question is contained in $G_1 \cap G_3$ without loss of generality. It is then easy to see that also some noncentral composition factor of the FF-module is again a FF-module for the same offending subgroup, and one determines without difficulty that this composition factor can be at most 4-dimensional, hence has to be a natural $SL_4(3)$-module for G_1/M_1 resp. G_3/M_3.

Now we reach a contradiction by constructing a nontrivial subgroup of B normalized by $G =< G_1, G_3 >$. Consider the group $X'_{2,4}$. By the structure of, say, $(G_1/M_1)'$, which is isomorphic to $SL_4(3)$ by (iii), there is an involution t in $B \cap X'_{2,4}$ that maps into the center of G_1/M_1 as well as of G_3/M_3. Hence the normal closure of $< t >$ under G is contained in B, and the desired contradiction follows.

3. List of examples

In this section, we want to list the various types of T-SCABs, and discuss whether there are finite examples known, whether the universal 2-cover is a building and so on.

We only list types with connected diagrams of rank at least three. We will give the diagram, the characteristic, the relevant parabolics and a finite example, if one is known. If there is a bunch of T-SCABs differing only very slightly (for instance only by a factor 2 on all the parabolics, and having isomorphic universal 2-covers), we give only one finite example. "Relevant" parabolics will be the ones with connected spherical diagram of maximal rank; sometimes we also give the parabolic isomorphism type of the stabilizer of a nonspherical residue if it appears already in the list, referring to its number in the list. Thus not always all rank 2 parabolics can be derived from the given data, and so sometimes, for instance if the diagram is  there are more than one parabolic isomorphism types for the given example. The possible types can, however, be easily obtained from the given information.

To make the table better to read and not to have too many entries in each row, we give references to the papers where the examples first occurred, where they were described and/or classified at the end of the section.

[A] Δ is spherical, C is a finite building and G is an extension of a finite simple Lie type group by diagonal and field automorphisms, or G is A_7 and C is the $L_4(2)$- building.

[B] Δ is affine, the universal 2-cover of C is the affine building of a simple algebraic group over a locally compact local field.

(1)		$p=2$	$\Omega_8^+(2)$	$\Omega_8^+(3)$
(2)	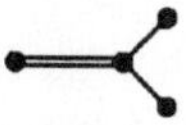	$p=2$	$Sp_6(2), Sp_6(2), 2^6A_7$	$\Omega_7(3)$
(3)		$p=2$	A_7	$\Omega_6^-(3)$
(4)		$p=2$	2^4X or $2^5X, X \in \{A_6, \Sigma_6\}$	$\Omega_6^-(3)$
(5)		$p=2$	A_6 or Σ_6	3^4A_6
(6)		$p=2$	$\Omega_6^-(2)$ or $O_6^-(2)$	$3^5\Omega_6^-(2)$
(7)		$p=2$	F_{21}	$L_3(2)$,
				F_{21}, tight
(8)		$p=2$	$F_{73.9}$	$F_{73\cdot 9}$, tight
(9)	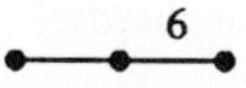	$p=2$	$2^3L_3(2), G_2(2)$	$G_2(3)$
(10)		$p=3$	$\Omega_5(3), SO_5(3)$ or	$2^{14}.\Omega_5(3)$
			$2^4X, 2^5X, X = A_5, \Sigma_5$ or F_{20}	
(11)		$p=3$	$2^4.X, X = A_5, \Sigma_5$ or F_{20}	$\Omega_5(5)$

[C] "Sporadic examples" (the universal 2-cover is a building, hence infinite, but only essentially one finite projection is known).

– $O_{p'}(X_{i,j}) = 1$ for all i, j (except in (3)) –

(1)		$p = 2$	$2^{1+6}\Omega_6^-(2), 2^{4+6}3A_6$	Suz
(2)	 6 6	$p = 2$	$G_2(2)$	$G_2(3)$
(3)		$p = 2$	F_{21}	A_7
(4)		$p = 3$	$3SO_5(3)$	$\Omega_8^+(2)$
(5)		$p = 3$	$U_4(3).2$	$U_6(2)$
(6)	 6	$p = 5$	$5^3SL_3(5), G_2(5)$	Ly
(7)	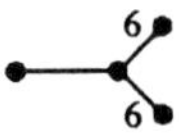 6 6	$p = 2$	B.8 and C.2, tight	$G_2(3)$
(8)		$p = 3$	C.4, tight	$\Omega_8^+(2)$
(9)		$p = 3$	C.5, tight	$U_6(2)$

[D] Universal 2-cover is a building, finite projections are known (though none given below).

– $O_p'(X_{i,j}) \neq 1$ for some i, j –

(1)		$p = 2$	B.2
(2)	$Y(r, s), r \geq 1, s \geq 2$	$p = 2, 3$ or 7	$PSp_{2n}(7)$, $PSp_{2n}(3), \Omega_{2n+1}(3), U_{2n}(3)$, $Sp_{2n}(2)$, $U_{2n}(2)$, $U_{2n+1}(2)$, (extended by diagonal and field automorphisms)
(3)	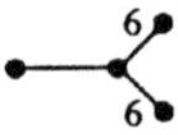 6 6	$p = 2$	B.9,C.2
(4)	complete graph only single bonds	$p = 2$	F_{21}, $L_n(2)$
(5)	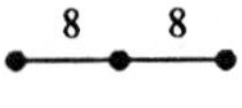 8 8	$p = 2$	$F_4(2)$ or $F_4(2)'$
(6)	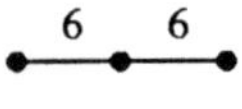 6 6	$p = 2$	$G_2(2), G_2(2)'$,

			${}^3D_4(2), Aut({}^3D_4(2))$
(7)	6 6	$p = 3$	$G_2(3), {}^3D_4(3), Aut({}^3D_4(3))$
(8)		$p = 3$	$L_3(4).2$ or $Aut(L_3(4))$
(9)		$p = 3$	$2^4X, X \in \{A_5, \Sigma_5, F_{20}\}$
(10)	complete bipartite graph, only double, triple, and quadruple bonds (all relevant rank 2 parabolics essentially isomorphic; in some cases only stars are possible)	$p = 2,3$	B.5,B.10,B.11, C.2,C.4,C.5,D.2 ($r = 1$) D.5,D.6,D.7,D.8,D.9
(11)	tripartite graph on $\{1\}\dot{\cup}I_1\dot{\cup}I_2$, only double bonds	$p = 2$	$X_{1,i,j}$ of type C.1 for $i \in I_1, j \in I_2$

[E] Universal cover is a building, in all cases there are local isomorphism types for which no finite projection is known (existence by Tits' theorem [35]).

(1)	complete graph only single bonds	$p = 2$	F_{21}
(2)	complete graph only single bonds	$p = 2$	$F_{73 \cdot 9}$
(3)	bipartite graph only double bonds,	$p = 3$	on $\{i, j\}$–residues with connected diagram, 2^4X, $X \in \{A_5, \Sigma_5, F_{20}\}$, is induced with kernel a $\{2,3\}$-group centralized by $O^{2,3}(X_{i,j})$.

[F] C_7 is residue, hence $p = 2$, universal 2-cover is not a building, tight extensions always contain C_7^*, notation as in (2.9). In many cases information on the chamber system rather than the residues is given.

(1)		$C = C_7$	$G = A_7$
		$C = C_7^*$	$G = A_7$

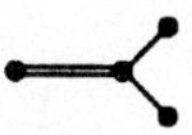

(2)		tight extension of D.1	
(3)		residue of F.2	
(4)		A_7, A_7, B.5, A_7	$\Omega_6^-(3)$
(5)		A_7, A_7, B.5, B.5	$\Omega_6^-(3)$
(6)		O_{12345}, tight extension of F.4,B.3	$\Omega_6^-(3)$
(7)		O_{12346}, tight extension of F.4, F.5	$\Omega_6^-(3)$
(8)		O_{123456}, tight extension of F.6, F.7	$\Omega_6^-(3)$
(9)		A_7, A_7, A_7, A_7	$U_3(5)$
(10)		A_7, A_7, A_7, A_7	$U_3(5)$
(11)		U_{12345}, tight extension of F.9, F.10	$U_3(5)$
(12)	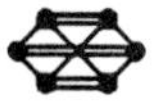	U_{123456}, tight extension of F.11	$U_3(5)$
(13)		$\widehat{2A_8}, 2^4A_7, 2^4A_7$, B.4	McL

The examples were found by many authors. Most of them were already listed and referenced in [6]. The references we give concern either the existence of a single finite example or existence and description of the universal 2-cover, hence an infinity of finite examples.

No comment is necessary for the examples in section A.

B.1, B.2, B.4 and B.9 appear in [7] and [2]; B.3 in [8] and [2]; B.4 in [5], B.5 in [15], B.6 and B.8 in [13], B.7 in [11] and [12], B.10 and B.11 in [10] and [13].

C.1 appears in [19], C.2 in [2], C.3 in [18], C.4, C.5, C.6 in [5], C.7 in [2], C.8 and C.9 are read off from [5].

D.4 was constructed in [36], D.2 appears in [2], the remaining cases of D, mainly "module extensions", in [13] and [3].

F.9 appears in [16], F.10, F.11, F.12 in [36], F.4, F.5, F.6, F.7, F.8 in [15] and F.13 in [20].

4. Remarks, open problems

In the list in section 3, some entries have to be read carefully; for instance for diagrams •—m—•—m—• , the isomorphism type of $X_{1,3}$ might vary, while the isomorphism types of $X_i, X_{1,2}$ and $X_{2,3}$ stay the same. But even the rank 1 parabolics in the group X_{12} might appear as X_1 or X_2. Hence some entries stand for quite a number of parabolic isomorphism types. On the other hand, the same parabolic isomorphism type does sometimes appear in more than one place. In these cases, additional information on the local isomorphism types was available.

As pointed out in the introduction, the classification up to parabolic isomorphism does not quite give a list of all T-SCABs. Some little extra work is needed to get a full local classification of T-SCABs.

One very interesting question about T-SCABs is: what can be said about the universal 2-covers? In particular in the cases where one knows that the universal 2-covers are buildings ([34]), one would like to know whether some "known" buildings arise as 2-covers. This question is answered by the Kantor-Liebler-Tits Theorem ([9]), which assures that the only "classical" (affine) buildings that are T-SCABs, i.e. for which the corresponding algebraic group over the local field possesses a discrete subgroup acting chamber transitively on the affine building, are the ones in B. But what can be said about the universal 2-covers in the remaining cases?

If some rank 3 residue is C_7, the universal 2-cover is not a building. What can be said about these simply 2-connected T-SCABs, and are some of them finite?

A positive answer to this question was obtained in several cases, for instance F.1,F.9,F.10,F.13 by Ronan resp. Stroth. Others, however, such as F.3 or F.4 are known to have infinite universal covers and to be strongly related to some building (Li, see [6]).

Problems: (a) *Determine for all types in F whether the universal 2-covers are finite or infinite.*

(b) *Is there a classification of tight T-SCABs – at least of tight rank n T-SCABs, where all rank $n-1$ residues are not tight? This could be interesting in view of some of the "sporadic" examples listed.*

References:

1. Aschbacher,M., Finite geometries of type C_3 with flag-transitive groups. Geom. Ded. 16 (1984), 195-200.
2. Aschbacher,M. and Smith,St., Tits geometries over GF(2) defined by groups over GF(3). Comm. Alg. 11 (1983), 1675-1684.

3. Cuypers,H. and Meixner,Th., Some extensions of dual polar spaces. Preprint.
4. Heiss,St., Extensions of the chamber system of type A_3 for A_7. Preprint FU Berlin 1988.
5. Kantor,W., Some geometries that are almost buildings. Europ. J. Comb. 2 (1981), 239-247.
6. Kantor,W., Generalized polygons, SCABs and GABs Proc. CIME-Conf. Buildings and the Geometry of Diagrams, Como 1984, Springer Lecture Note 1181, 79-158.
7. Kantor,W., Some exceptional 2-adic buildings. J. of Alg. 92 (1985), 208-223.
8. Kantor,W., Some locally finite flag-transitive buildings. Europ. J. Comb. 8 (1987) 429-436.
9. Kantor,W. ,Liebler,R. and Tits,J., On discrete chamber-transitive automorphism groups of affine buildings. Bull.AMS 16 (1987), 129-133.
10. Kantor,W., Meixner,Th. and Wester,M., Two exceptional 3-adic affine buildings. To appear in Geom. Ded.
11. Köhler,P., Meixner,Th. and Wester,M., The 2-adic affine building of type $\tilde{A}_2$ and its finite projections. JCT(A) 38 (1985), 203-209.
12. Köhler,P., Meixner,Th. and Wester,M., The affine building of type $\tilde{A}_2$ over a local field of characteristic two. Arch. Math. 42 (1984), 400-407.
13. Meixner,Th., Klassische Tits Kammersysteme mit einer transitiven Automorphismengruppe. Mitt. Math. Sem. Giessen 174 (1986).
14. Meixner,Th., Locally finite classical Tits chamber systems with transitive group of automorphisms in characteristic 3. To appear in Geom. Ded.
15. Meixner,Th. and Wester,M., Some locally finite buildings derived from Kantor's 2-adic groups. Comm. Alg. 14 (1986), 389-410.
16. Neumaier,A., Some sporadic geometries related to PG(3,2). Arch. Math. 42 (1984), 89-96.
17. Niles,R., Finite groups with parabolic type subgroups must have BN-pair. J. Alg. 75 (1982), 484-494.
18. Ronan,M., Triangle geometries. JCT(A) 37 (1984), 294-319.
19. Ronan,M. and Smith,St., 2-local geometries for some sporadic groups. Proc. Symp. Pure Math. 37 (1980), 283-289.
20. Ronan,M. and Stroth,G., Minimal parabolic geometries for the sporadic groups. Eur. J. Comb. 5 (1984), 59-92.
21. Seitz,G., Flag-transitive subgroups of Chevalley groups. Ann. of Math. 97 (1973), 27-56 (correction unpublished).

22. Stroth,G., Geometries of type M related to A_7. Geom. Ded. 20 (1986), 265-293.

23. Stroth,G., Chamber Systems, geometries and parabolic systems whose diagram contains only bonds of strength 1 and 2. Preprint FU Berlin (1985).

24. Stroth,G., A local classification of finite classical Tits geometries of characteristic $\neq 3$. Geom. Ded. 28 (1988) 93-106.

25. Thiel,H., (unpublished).

26. Timmesfeld,F., Tits Geometries and Parabolic systems in Finitely Generated Groups I, II. Math.Z.184 (1983), 377-396, 449-487.

27. Timmesfeld,F., Locally finite classical Tits chamber systems of large order. Invent.math. 87 (1987), 603-641.

28. Timmesfeld,F., Tits Chamber Systems and Finite Group Theory. Proc.CIME-Conf. Buildings and the Geometry of Diagrams, Como 1984, Springer Lecture Note 1181, 249-269.

29. Timmesfeld,F., Classical Locally Finite Tits Chamber Systems of Rank 3 and Characteristic 2. Preprint.

30. Timmesfeld,F., Tits geometries and revisionism of the classification of finite simple groups of characteristic 2 type. Proc. Rutgers Group Theor. Year 83-84, Cambridge Univ. Press (1984), 229-242.

31. Timmesfeld,F., Amalgams with Rank 2 Groups of Lie-type in Characteristic 2. Preprint.

32. Timmesfeld,F., Classical Locally Finite Tits Chamber Systems of Rank 3. J. of Algebra 124 (1989) 9-59.

33. Tits,J., Buildings of Spherical Type and Finite BN-Pairs. Springer Lecture Note 386, (1974).

34. Tits,J., A local approach to buildings. In: The Geometric Vein (Coxeter Festschrift) 519-547, Springer (1982).

35. Tits,J., Buildings and group amalgamations. Proc. Groups - St. Andrews 1985, LMS Lecture Notes 121 (1986), 110-127.

36. Wester,M., Endliche fahnentransitive Tits Geometrien und ihre universellen Überlagerungen. Mitt. Math. Sem. Giessen 170 (1985).

Representation Theory of Chambersystems of Affine Type

Udo Ott

For more than ten years geometry has been in an important period dominated by ideas developed in the course of laying the foundation for Lie geometries. Recent work on certain classification problems for geometries of spherical type point to the significance of representation theory for the analysis of these geometries. The general method of using Hecke algebras provides the appropriate tool for studying certain aspects of the subject.

The question arises whether the Hecke algebra will continue to be useful in problems involving chambersystems of affine type. The purpose of this summary is to give a report on the program to find a workable description of the finite dimensional irreducible representations of the affine Hecke algebras.

Remarkably enough, at this early stage representation theory enables us to introduce the notion of dimension for affine chambersystems.

1. Chambersystems and Hecke algebras

We use throughout the paper the notation and terminology of (8, 9, 11). Let M be a Coxeter diagram over the set $I = \{0, 1, \ldots, n\}$ and let $\mathcal{C} = (C, \mathcal{P}_0, \mathcal{P}_1, \ldots, \mathcal{P}_n)$ be a chambersystem with parameters $q_0, q_1, \ldots, q_n$ admitting M as a diagram in the sense of Buekenhout and Tits (11).

It is convenient to write

$$\Delta \underset{i}{\sim} \Gamma$$

for two different chambers Δ and Γ in a class of the partition. In particular, we associate with the chambersystem the colored graph

$$\mathcal{C}^* = (C, \underset{0}{\sim}, \underset{1}{\sim}, \ldots, \underset{n}{\sim}).$$

Our assumption then implies that for $i, j \in M$ and $i \neq j$ the connected components of the subgraph $(C, \underset{i}{\sim}, \underset{j}{\sim})$ are the chambersystems of generalized $M(i,j)$-gons with parameters q_i, q_j.

We assume also that the chambersystem is connected, this means that every two chambers are connected by a gallery. Let $W = \langle r_0, r_1, \ldots, r_n \rangle$ be the corresponding Coxeter group and let $H = \langle \sigma_0, \sigma_1, \ldots, \sigma_n \rangle$ be the Hecke algebra of the chambersystem defined over the field of complex numbers.

We recall that the standard module V is the space of all maps with finite support from the set of chambers into the field of complex numbers.

For each word $\omega = i_1 i_2 \dots i_s$ in the free monoid F over the set I, the elements r_ω and σ_ω are defined by

$$r_\omega = r_{i_1} \dots r_{i_s} \text{ and } \sigma_\omega = \sigma_{i_s} \dots \sigma_{i_1}.$$

It is immediate that $r_\omega r_\tau = r_{\omega\tau}$ and $\sigma_\omega \sigma_\tau = \sigma_{\tau\omega}$ for $\omega, \tau \in F$. With the notation in (11, 3.1) we consider the word $p(i,j) = \dots jiji \dots jji$ of length $M(i,j)$ and obtain the following relations

$$\begin{aligned} &r_i^2 = 1 \\ (1) \qquad &r_{p(i,j)} = r_{p(j,i)} \quad i,j \quad i \neq j \end{aligned}$$

and

$$\begin{aligned} &\sigma_i^2 = q_i \cdot 1 + (q_i - 1)\sigma_i \\ (2) \qquad &\sigma_{p(i,j)} = \sigma_{p(j,i)} \quad i,j \quad i \neq j. \end{aligned}$$

If ω and τ are homotopic words, we write $\omega \simeq \tau$. By (1) and (2),

$$\sigma_\omega = \sigma_\tau \quad \text{if} \quad \omega \simeq \tau.$$

For each element $w = r_{i_1} r_{i_2} \dots r_{i_s}$ in the Coxeter group W let

$$\lambda(w) = \sigma_{i_1} \sigma_{i_2} \dots \sigma_{i_s}$$

assuming that $r_{i_1} r_{i_2} \dots r_{i_s}$ is a reduced presentation of w. By well known properties of Coxeter groups, this definition makes sense. We introduce also a cover $H(M)$ of the Hecke algebra, which has generators $\Sigma_0, \Sigma_1, \dots, \Sigma_n$ and defining relations (2) with Σ_i instead of σ_i. Keeping the above notation, we note that the element

$$\Lambda(w) = \Sigma_{i_1} \dots \Sigma_{i_s}$$

is well defined. Moreover, it follows at once from the definition that the standard module becomes a left $H(M)$-module in a natural way.

So far we have introduced the general setting. In the following we assume that the chambersystem is of affine type $\tilde{X}$, where X denotes the spherical part. Thus M is one of the following diagrams:

$\tilde{A}_1$ ∞ ; 0

$\tilde{A}_n \quad (n \geq 2)$ 0

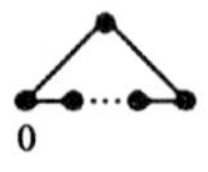

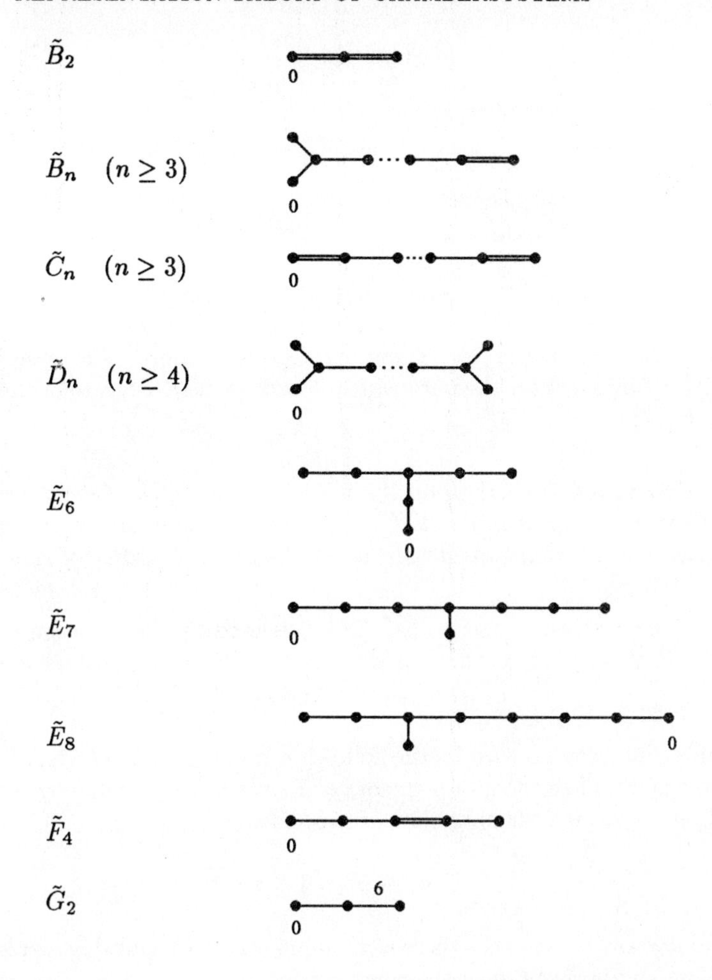

We remark that from our algebraic point of view also some "sporadic" diagrams for geometries lead to diagrams of spherical or affine type (compare 8). For instance we have diagrams attached to certain groups:

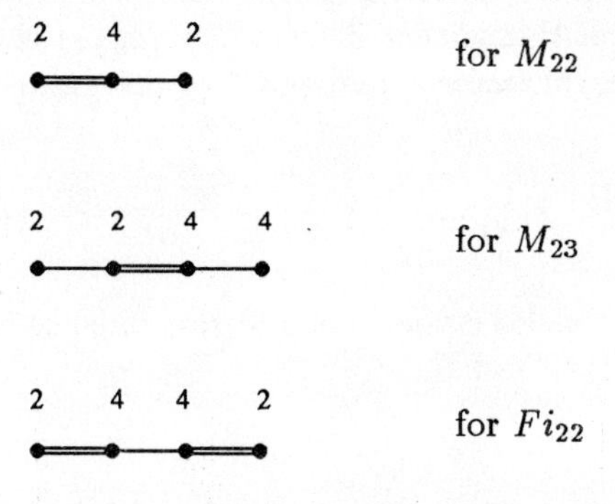

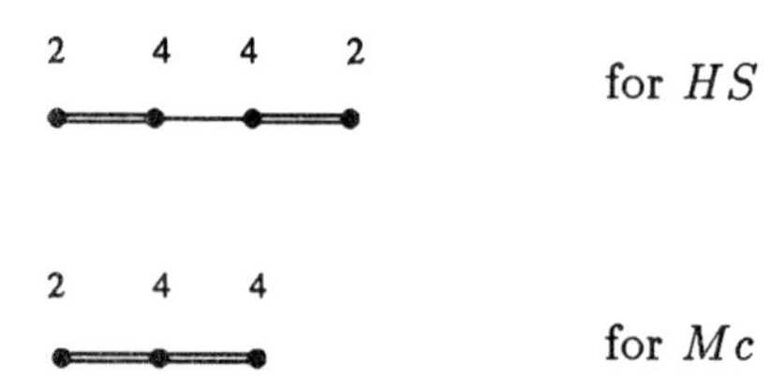

The structure of the cover $H(M)$ appears intimately connected with the structure of the corresponding Coxeter group. Therefore we begin with a fundamental theorem which describes the structure of the group W (comp. 1).

Theorem 1: *The subgroup $W_0 = \langle r_1, \ldots, r_n \rangle$ of the Coxeter group W is a Coxeter group of spherical type X and admits an abelian subgroup T of W as normal complement. Moreover the group T is free of rank n.*

The algebraic variety $\mathcal{H} \subseteq \mathbf{C}^{n+1}$, defined by the equation

$$X_{n+1} = X_1 X_2 \ldots X_n$$

comes into the picture in the following manner. Let $\{t_1, t_2, \ldots, t_n\}$ be a basis of T. Then the group algebra of T over $\mathbf{C}$ is generated by the elements $t_1, t_2, \ldots, t_n$ and we have the correspondence

$$t_i \leftrightarrow X_i \quad 1 \leq i \leq n,$$

and it is easily seen that there is a unique way to extend this correspondence to an isomorphism with the ring

$$R = \mathbf{C}[X_1, \ldots, X_n, 1/X_1 X_2 \ldots X_n],$$

which is the ring of regular functions of the mentioned variety. One interesting consequence of this is that if $(z_1, \ldots, z_n, z_{n+1})$ is a point of the variety, then we find a character $\theta : T \to \mathbf{C}^*$ of the group satisfying the equation

$$\theta(t_i) = z_i \quad 1 \leq i \leq n.$$

In fact this formula yields a one-to-one correspondence between $\mathcal{H}$ and the dual group $\mathrm{HOM}(T, \mathbf{C}^*)$ of T.

2. The Generic Representation

Let H_0 be the subalgebra of $H(M)$ generated by the elements

$$\Sigma_1, \Sigma_2, \ldots, \Sigma_n.$$

To prove that the relations given in (2) among the generators $\Sigma_1, \Sigma_2, \ldots, \Sigma_n$ are defining relations for H_0, it is obviously necessary and sufficient to show that the elements $\Lambda(w)$, $w \in W_0$ are linearly independent. By introducing a certain $H(M)$-module one can even prove that (compare for instance 6)

$$\{\Lambda(w) \mid w \in W\} \quad \text{is a basis of } H(M). \tag{3}$$

Thus we have

$$\{\Lambda(w) \mid w \in W_0\} \quad \text{is a basis of } H_0. \tag{4}$$

We denote by V_0 the regular left H_0-module given by the algebra itself. The dimension is

$$\dim_{\mathbf{C}} V_0 = |W_0|.$$

Our program is to regard V_0 as an $H(M)$-module. Thus we are led to the interesting problem to find admissible actions for the generator Σ_0 on V_0.

This is possible but by no means obvious. The way to describe these actions is to construct a generic representation of $H(M)$ over the ring R of regular functions of $\mathcal{H}$. Thus, relative to the basis $\{\Lambda(w) \mid w \in W_0\}$ the action of Σ_0 is given by an admissible generic $|W_0| \times |W_0|$-matrix over R. To be precise, we introduce the extended algebra and module

$$H_0^R = H_0 \otimes_{\mathbf{C}} R = V_0^R.$$

We form also

$$H_0^K = H_0 \otimes_{\mathbf{C}} K = V_0^K$$

where K denotes the field of rational functions on $\mathcal{H}$. It follows at once from the definitions that V_0^K becomes a left H_0^K-module of dimension $|W_0|$ over K. Furthermore, we may regard V_0^R as an H_0^R-invariant lattice of rank $|W_0|$ in V_0^K. Then $\{\Lambda(w) \mid w \in W_0\}$ is a basis of V_0, V_0^R, V_0^K over $\mathbf{C}, R, K$.

We say that a K-linear map

$$\Sigma_0^* : V_0^K \to V_0^K$$

is an *admissible* map provided the following conditions hold:

(a) $\Sigma_0^* V_0^R \subseteq V_0^R$

(b) The K-algebra $\mathrm{End}_K(V_0^K)$ is generated by the elements $\Sigma_0^*, \Sigma_1, \Sigma_2, \ldots, \Sigma_n$.

(c) The elements $\Sigma_0^*, \Sigma_1, \Sigma_2, \dots, \Sigma_n$ satisfy the relation (2).

The important point about (c) is that the correspondence

$$\begin{aligned} \Sigma_0 &\leftrightarrow \Sigma_0^* \\ \Sigma_i &\leftrightarrow \Sigma_i \quad 1 \leq i \leq n \end{aligned}$$

makes V_0^R into an $H(M)$-module.

Furthermore, we obtain a series $V(z)$, $z \in \mathcal{H}$ of $H(M)$-modules in the following manner. For each point $z \in \mathcal{H}$ the specialization $\phi(z) : V_0^R \to V_0$ (evaluation at z) yields an $H(M)$-module $V(z)$, which is as an H_0-module isomorphic to the module V_0. One easily checks that only the element $\alpha = 0$ of $H(M)$ satisfies the equation $\alpha V(z) = 0$ for all $z \in \mathcal{H}$. Thus a theorem of Kaplansky (comp. 2) implies

Theorem 2: *Assume that there is an admissible map. Then a finite dimensional irreducible $H(M)$-module is at most $|W_0|$-dimensional.*

We also conclude that our algebra $H(M)$ can be presented as an algebra of $|W_0| \times |W_0|$-matrices over R:

Theorem 3: *Assume that there is an admissible map. Then the algebra $H(M)$ is isomorphic to the* **C***-algebra $A \subseteq \mathrm{End}_K(V_0^K)$ generated by the elements $\Sigma_0^*, \Sigma_1, \Sigma_2, \dots, \Sigma_n$. Furthermore, we have the isomorphism $A \otimes_{\mathbf{C}} K \cong \mathrm{End}_K(V_0^K)$.*

We state the main theorem of (5) which rests on basic results of (7):

Theorem 4: *There are admissible maps with the property such that a finite dimensional irreducible $H(M)$-module is isomorphic to a composition factor of a specialization $V(z)$ for some point z of $\mathcal{H}$.*

Example. Since the Hecke algebra of a finite regular geometry admits the cover $H(\tilde{A}_1)$, we obtain the known result that the Hecke algebra has at most 2-dimensional irreducible modules (comp. 3). With respect to the basis $\Lambda(1)$, $\Lambda(r_1)$ a generic matrix is

$$\begin{pmatrix} q_0 - 1 & 1/X \\ q_0 X & 0 \end{pmatrix}.$$

Thus the algebra $H(\tilde{A}_1)$ is isomorphic to the **C**-algebra generated by the two matrices

$$\begin{pmatrix} 0 & q_1 \\ 1 & q_1 - 1 \end{pmatrix}, \begin{pmatrix} q_0 - 1 & 1/X \\ q_0 X & 0 \end{pmatrix}.$$

3. Dimension of the Chambersystem

A considerable number of chambersystems of affine type have been constructed. Usually they are covered by buildings (comp. 11). But apart from this result the current situation seems to be of an enormous complexity.

The following notion of the dimension of a chambersystem may throw some light into the situation. The *spectrum* of the chambersystem $\mathcal{C}$, denoted by Spec($\mathcal{C}$), is the set of points $z \in \mathcal{H}$ such that a composition factor of $V(z)$ is isomorphic to an irreducible factor of the standard module V of the chambersystem. With respect to the Zariski topology the algebraic closure of the spectrum is a subvariety $\mathcal{H}(\mathcal{C})$ of $\mathcal{H}$. The algebraic dimension of $\mathcal{H}(\mathcal{C})$ is the dimension of the chambersystem, denoted by $\dim(\mathcal{C})$.

The following result should not be a surprise (5):

Theorem 5: *Let $\mathcal{C}$ be a chambersystem of the affine type M.*
(a) *If $\mathcal{C}$ is finite then* $\dim(\mathcal{C}) = 0$.
(b) *If $\mathcal{C}$ is a building then* $\dim(\mathcal{C}) = n$.

Obviously it would be desirable to obtain a characterization theorem for buildings which depends upon the notion of dimension.

References:

(1) N. Bourbaki, Groupes et algèbres de Lie. Hermann Paris, 1968.

(2) R. Godement, A theory of spherical functions, Trans. Amer. Math. Soc. 73, pp. 496-556, 1952.

(3) R. Liebler, Tactical configurations and their generic ring, preprint.

(4) ________, A representation theoretic approach to finite geometries of spherical type, preprint.

(5) S. Löwe, Über die Struktur der Hecke Algebren vom affinen Typ, Dissertation TU Braunschweig, 1988.

(6) K. Lusztig, Hecke algebras, Lecture Notes of the University of Rome.

(7) H. Matsumoto, Analyse Harmonique dans les Systèmes de Tits Bornologiques de Type Affine, Lecture Notes in Mathematics 590, Springer Berlin.

(8) U. Ott, Bericht über Hecke Algebren und Coxeter Algebren endlicher Geometrien, in "Finite Geometries and Designs", London Math. Soc. Lecture Note Series 49, pp. 260-271, 1981.

(9) ________, Some remarks on representation theory in finite geometry, in Lecture Notes in Mathematics 893, pp. 68-110, Springer Berlin.

(10) ________, On chambersystems of type $\tilde{A}_2$, Lecture Note 38, University of Naples.

(11) J. Tits, A local approach to buildings, in "The Geometric Vein", pp. 519-547, Springer Berlin, 1981.

The geometry of trilinear forms

Michael Aschbacher*

All of us at this conference are familar with projective geometry on a vector space V over a field F, and the refinements of projective geometry induced by a bilinear form on V. This note discusses the analogous geometries induced by trilinear forms on V.

There does not seem to be much in the literature on this subject. In [4], Cohen and Helminck determine all alternating trilinear forms of dimension at most 7 over fields of cohomological dimension at most 1. In [1], [2], [3] there is some discussion of alternating and symmetric trilinear forms. There is of course an extensive literature on Lie algebras and Jordan algebras which is of some relevance.

In the case of bilinear forms, the study of alternating forms is easier than the study of symmetric forms. The same seems to be true of trilinear forms. Nevertheless I will restrict my discussion to symmetric trilinear forms, as I have done more work with symmetric forms than alternating forms. However much of the discussion carries over to alternating forms.

I have no strong results to offer. Section 1 describes some examples. Sections 2 and 3 develop some elementary concepts. Section 4 poses a few problems. Much of the discussion is motivated by attempts to understand the subgroup structure of E_6 via the E_6-invariant 27-dimensional symmetric trilinear form.

Throughout this paper, F is a field and V is a finite dimensional vector space over F.

1. Examples

Write $SY^k(V)$ for the space of symmetric k-linear forms on V. Given $f \in SY^3(V)$ and a basis X of V write

$$f = \sum_{(x,y,z)\in X^3} a_{x,y,z}xyz$$

to indicate that $f(x,y,z) = a_{x,y,z}$. The term $a_{x,y,z}xyz$ is a *monomial* of f. We exhibit some examples of symmetric trilinear forms.

* Partially supported by NSF DMS-8721480 and NSA MDA90-88-H-2032.

(1) *The E_6-form.*

Let V be of dimension 27 with basis

$$X = (x_i, x'_i, x_{ij} : 1 \leq i, j \leq 6, \; i \neq j)$$

subject to the convention that $x_{ji} = -x_{ij}$. Let f be the symmetric trilinear form on V with monomials

$$x_i x'_j x_{ij}, \; 1 \leq i, j \leq 6, \; i \neq j,$$

$$x_{1d,2d} x_{3d,4d} x_{5d,6d}, \; d \in Coset$$

where $Coset$ is some set of coset representatives for Alt_P in Alt, Alt is the alternating group on $\{1, ..., 6\}$, and Alt_P is the stabilizer in Alt of the partition $P = \{\{1,2\},\{3,4\},\{5,6\}\}$ of $\{1, ..., 6\}$. The group of isometries of f is the universal Chevalley group $E_6(F)$ over F. This form appears in Cartan [5] and Dickson [6].

(2) *The 15-dimensional L_6-form.*

Let U be a 6-dimensional space over F and identify F with $\Lambda^6(U)$. Let $V = \Lambda^2(U)$ and define $f \in SY^2(V)$ by

$$f(x \wedge y, u \wedge v, z \wedge w) = x \wedge y \wedge u \wedge v \wedge z \wedge w, \; x, y, u, v, z, w \in U.$$

Observe that $SL(U)$ is represented as a group of isometries of f via its action on V. Also the 15-dimensional L_6-form is isometric to the restriction of the E_6-form to the subspace $V_{15} = \langle x_{ij} : i \neq j \rangle$.

(3) *The 9-dimensional tensor product form.*

Let A, B be 3-dimensional vector spaces over F. Then $SL(A)$ preserves the alternating trilinear form $f_A = x_1 x_2 x_3$ on A, where $\{x_1, x_2, x_3\}$ is a basis for A. Similarly $SL(B)$ preserves the alternating form $f_B = y_1 y_2 y_3$. Let $V = A \otimes B$ and $f = f_A \otimes f_B$. That is

$$f(a_1 \otimes b_1, a_2 \otimes b_2, a_3 \otimes b_3) = f_A(a_1, a_2, a_3) f_B(b_1, b_2, b_3).$$

Then f is a symmetric trilinear form on V and $SL(A) \times SL(B)$ is represented as a group of isometries of f via the tensor product action on V. Observe the 9-dimensional tensor product form is isometric to the restriction of the E_6-form to the subspace

$$V_9 = \langle x_{ij} : 1 \leq i \leq 3, \; 4 \leq j \leq 6 \rangle.$$

(4) *The 6-dimensional L_3-form.*

In the previous example take $char(F) \neq 2$, $B = A$, and let V be the subspace of $A \otimes A$ generated by $x \otimes x$, $x \in A$. Now $SL(A)$ acts on $A \otimes A$

via the diagonal action with V an $SL(A)$-submodule of $A \otimes A$ isomorphic to $S^2(A)$. Let f be the restriction of the tensor product form to V.

If $f \in SY^3(V)$ and $\langle\ ,\ \rangle$ is a nondegenerate symmetric bilinear form on V, then f and $\langle\ ,\ \rangle$ induce a Norton algebra structure on V. Namely we define a product $(x,y) \mapsto xy$ on V by letting xy be the unique element of V such that $f(x,y,z) = \langle xy, z\rangle$ for all $z \in V$. This product defines a commmutative nonassociative algebra structure on V preserved by the isometry group of the pair $f, \langle\ ,\ \rangle$. If f is the E_6-form and we choose $\langle\ ,\ \rangle$ judiciously, our Norton algebra is a Jordan algebra J with automorphism group $F_4(F)$. Moreover each subspace U in examples (1)-(4) is a Jordan subalgebra J_U of J. However in each case we have exhibited a group L_U of isometries of f_U such that U is not selfdual as an FL_U-module, so L_U does not preserve $\langle\ ,\ \rangle$ and indeed does not preserve J_U. That is $Aut(f_U)$ is bigger than $Aut(J_U)$.

(5) *The Griess algebra.*
Griess constructed a 196,884-dimensional complex algebra A and showed $Aut(A)$ is the Monster. The Griess algebra is the Norton algebra for a suitable $Aut(A)$-invariant pair $f, \langle\ ,\ \rangle$ of forms.

Let $F = GF(2)$ and $P : V \to F$ a function with $P(0) = 0$. For m a positive integer and $x = (x_1, \ldots, x_m) \in V^m$, write 2^m for the power set of $\{1, ..., m\}$ and for $I \in 2^m$ let $I(x) = \sum_{i \in I} x_i$. Define $P_m : V^m \to F$ by $P_m(x) = \sum_{I \in 2^m} P(I(x))$. Define $deg(P)$ to be the maximum d such that $P_d \neq 0$. Notice that if $\{x_1, \ldots, x_m\}$ is linearly dependent then $P_m(x) = 0$. Thus $deg(P)$ is well defined and $deg(P) \leq dim(V)$. Observe also that if $d = deg(P)$ then $P_d \in SY^d(V)$. For each P_m is symmetric with $P_m(x+y, x_2, \ldots, x_m)$ equal to

$$P_m(x, x_2, \ldots x_m) + P_m(y, x_2, \ldots, x_m) + P_{m+1}(x, y, x_2, \ldots, x_m).$$

Next suppose (V, X) is a binary error correcting code; that is X is a basis for some F-space U and V is a subspace of U. Identify each $v = \sum_{x \in X} a_x x$ with its support $\{x \in X : a_x = 1\}$ and let $e = e(V, X)$ be maximal subject to $|v| \equiv 0 \bmod 2^e$ for all $v \in V$. Define P by $P(v) = |v|/2^e \bmod 2$. Then it is easy to check that $deg(P) = e + 1$. Indeed

$$P_m(v_1, \ldots, v_m) = |v_1 \cap \ldots \cap v_m|/2^{e+1-m} \bmod 2.$$

In particular if $e(V, X) = 2$ then P_3 is a symmetric trilinear form on V preserved by $Aut(V, X)$. Also if $z = \sum_{x \in X} x \in V$ and $|z| \equiv 0 \bmod 2^{e+1}$, then $z \in Rad(P_{e+1})$.

(6) *The* 11-*dimensional* M_{24}-*form.*

Let (V, X) be the 12-dimensional binary Golay code and $G = Aut(V, X) \cong M_{24}$. Then the weights of the vectors of V are 0, 8, 12, 16, 24, so $e(V, X) = 2$, $P_3 \in SY^3(V)$, and $z = \sum_{x \in X} x$ is in the radical of P_3. Thus P_3 induces $f \in SY^3(V/\langle z \rangle)$ and G is the isometry group of f.

2. Elementary algebra of trilinear forms

Let $G \leq GL(V)$. To simplify the discussion we assume $char(F) \neq 2$. This is not a serious constraint; modulo some minor modifications the discussion below goes through when F is of characteristic 2.

Let $L_G^3(V)$, $SY_G^3(V)$ be the set of G-invariant trilinear forms, symmetric trilinear forms on V, respectively. Given a trilinear form f on V and $x \in V$, define $f_x : V \times V \to F$ by $f_x(y, z) = f(x, y, z)$.

2.1. *Let S be the space of forms $f \in L_G^3(V)$ with $f(x, y, z) = f(x, z, y)$ for all $x, y, z \in V$. Then the map $\phi : S \to Hom_{FG}(V, SY^2(V))$ defined by $\phi(f) : x \mapsto f_x$ is an isomorphism. In partiuclar ϕ induces an injection of $SY_G^3(V)$ into $Hom_{FG}(V, SY^2(V))$.*

Proof: This is straightforward. Observe $\phi^{-1}(\alpha)(x, y, z) = \alpha(x)(y, z)$ for $\alpha \in Hom_{FG}(V, SY^2(V))$ and $x, y, z \in V$.

2.2. $SY^2(V) \cong S^2(V)^* \cong S^2(V^*)$.

Lemmas 2.1 and 2.2 tell us that the space of G-invariant symmetric trilinear forms on V is isomorphic to a subspace of $Hom_{FG}(V, S^2(V^*))$. In particular if this space is one dimensional then up to a scalar multiple there is a at most one nontrivial G-invariant symmetric trilinear form.

Let $\Delta(V, f)$ be the subgroup of $GL(V)$ consisting of those g such that for some $\lambda(g) \in F^\#$ and all $x, y, z \in V$, $f(xg, yg, zg) = \lambda(g) f(x, y, z)$. Thus $\Delta(V, f)$ is the group of *similarities* of f and the isometry group $O(V, f)$ consists of those $g \in \Delta(V, f)$ with $\lambda(g) = 1$. General nonsense tells us:

2.3. *Let $G \leq O(V, f)$ with G perfect, C the conjugacy class of G under $GL(V)$, and B the set of $b \in SY_G^3(V)$ similar to f. Then the number of orbits of $\Delta(V, f)$ on $O(V, f) \cap C$ is equal to the number of orbits of $N_{GL(V)}(G)$ on B.*

For example in theory (and sometimes in practice) we can use lemma 2.3 to determine the number of orbits of the similarity group of the E_6-form on subgroups of $E_6(F)$ isomorphic as linear groups to G.

3. Geometry of trilinear forms

Define a 3-*form* on V to be a triple $\mathcal{F} = (T, Q, f)$ such that $T : V \to F$, $Q : V^2 \to F$, $f \in SY^3(V)$, Q is linear in the first variable, and for all $x, y, z \in F$ and $a \in F$:

(a) $T(x+y) = T(x) + T(y) + Q(x,y) + Q(y,x)$.
(b) $T(ax) = a^3T(x)$.
(c) Q_x *is a quadratic form with associated bilinear form* f_x*; ie.*

$$Q_x(y+z) = Q_x(y) + Q_x(z) + f_x(y,z),$$

where $Q_x : y \mapsto Q(x,y)$.

Thus T is a cubic form and Q is a linear family of quadratic forms. If $char(F) \neq 2$ or 3 then $T(x) = f(x,x,x)/6$ and $Q(x,y) = f(x,y,y)/2$. If $char(F) = 2$ or 3, X is an ordered basis for V, and $f \in SY^3(V)$ with $f(x,x,y) = 0$ for all $x, y \in X$ then $\mathcal{F} = (T,Q,f)$ is a 3-form where

$$Q(v, \sum_{x \in X} a_x x) = \sum_{x<y} a_x a_y f(v,x,y),$$
$$T(\sum_x a_x x) = \sum_{x<y<z} a_x a_y a_z f(x,y,z).$$

Indeed $\mathcal{F}$ is unique subject to $T(x) = Q(v,x) = 0$ for all $v \in V$, $x \in X$. (*ie.* Each member of X is brilliant and singular in the terminology defined below.) So if $G \leq O(V,f)$ and for all $x \in X$, $u \in xG$, u is brilliant and singular, then $G \leq O(V,\mathcal{F})$. Thus if f is one of our first four examples then G preserves the unique 3-form $\mathcal{F} = (T,Q,f)$ making the standard basis X brilliant and singular. On the otherhand in example 6, M_{24} preserves no 3-form on V.

In the remainder of this section assume $\mathcal{F} = (T,Q,f)$ is a 3-form on V. Before continuing, let us recall the standard treatment of quadratic forms. If q is a nondegenerate quadratic form on V with bilinear form $\langle\ ,\ \rangle$ then:

(i) V *is the orthogonal direct sum of indecomposable subspaces* U *which cannot be written as the othogonal direct sum of proper subspaces. These indecomposables are of dimension* 1 *or* 2.
(ii) (V,q) *has the Witt property: If* $\alpha : (U,q) \to (W,q)$ *is an isomorphism of subspaces of* (V,q) *then* α *extends to an automorphism of* (V,q).

For $U \leq V$ define

$$U\theta = \{v \in V : (U,Q_v) \text{ is totally singular}\}.$$

As Q is linear in its first variable, $U\theta \leq V$. The analogue of the notion of 'orthogonal direct sum' for 3-forms is the operator $\oplus\theta$ defined by $V = A \oplus\theta\ B$ if $V = A \oplus B$, $A \leq B\theta$, and $B \leq A\theta$. We define V to be *indecomposable* if $V \neq A \oplus\theta\ B$ for proper subspaces A and B. Unfortunately the indecomposable summands in a decomposition of a space $(V,\mathcal{F})$

analogous to (i) need not be of small dimension. For example the spaces in examples (1)-(4) are indecomposable with respect to $\oplus\theta$ and 4.1 supplies indecomposables of arbitrarily high dimension. (*cf.* 3.4) Similarly (ii) fails miserably, as we will see in a moment. Thus the study of 3-forms is much more difficult than the study of quadratic forms.

Define $U \leq V$ to be *brilliant* if the restrictions of T and Q to U are trivial. This implies that f is also trivial on U. Define a point $\langle v\rangle$ of V to be *dark* if $\langle v\rangle$ is not brilliant. Define U to be *singular* if $V = U\theta$; equivalently, (U, Q_v) is totally singular for all $v \in V$. Observe that singular spaces are brilliant unless possibly $char(F) = 3$. However in general there are many nonsingular brilliant spaces. For example if f is the E_6-form then $V_6 = \langle x_1, \ldots, x_6\rangle$ is a maximal singular subspace and $V_{10} = \langle x_i, x_{i6} : 1 \leq i \leq 5\rangle$ is brilliant but nonsingular. This shows that the Witt property fails. However in examples (1)-(4) we have much symmetry. In particular in each case the similarity group is transitive on singular, nonsingular brilliant, and dark points.

If the isometry group of a 3-form is big then the form should possess singular points, as the following observation suggests:

3.1. *Let F be algebraically closed and G a closed connected semisimple subgroup of $O(V, \mathcal{F})$ irreducible on V. Let B be a Borel group of G and $\langle v\rangle \neq V$ a high weight vector for B. Then v is singular.*

Proof: Let H be a maximal torus of B and λ the weight of H on $\langle v\rangle$. If $T(v) \neq 0$ then for $h \in H$, $T(v) = T(vh) = T(\lambda(h)v) = \lambda(h)^3 T(v)$, so $3\lambda = 0$, a contradiction.

Similarly if $v\theta \neq V$ then $v\theta$ is a B-invariant hyperplane of V. Hence $V = v\theta \oplus \langle u\rangle$, where u is a low weight vector for H. Let μ be the weight of u. Then $0 \neq Q(u, v) = Q(uh, vh) = \mu(h)\lambda(h)^2 Q(u, v)$, so $\mu = -2\lambda$. By symmetry, $\lambda = -2\mu$, so $3\lambda = 0$, a contradiction.

For $u \in V$, define $u\Delta = Rad(f_u)$. For $U \subseteq V$, define $U\Delta = \bigcap_{u\in U} u\Delta$. Thus $U\Delta \leq V$. In example (1),

$$x_1\Delta = \langle x_1, x_1', x_i, x_{ij} : i, j \neq 1\rangle.$$

Notice the relation $U \leq W\Delta$ is symmetric. Also a subspace U of V is singular if and only if U is generated by a set S of singular points such that $t \in s\Delta$ for all $s, t \in S$. Indeed:

3.2. *Let A, B be distinct singular points of V. Then either*

(1) *$A \leq B\Delta$, $A + B$ is a singular line, and each point on $A + B$ is singular, or*

(2) *$A \not\leq B\Delta$, $A+B$ is nonsingular, A, B are the unique singular points on $A+B$, $(A+B)\theta$ is a hyperplane of V, and $(A+B, Q_v)$ is a hyperbolic line for each $v \in V-(A+B)\theta$.*

Call the lines in 3.2.2, ***hyperbolic***. Thus a pair of singular points generate a singular or hyperbolic line. Define a subspace U of V to be ***subhyperbolic*** if U is brilliant and $U\theta$ is a hyperplane of V. Thus hyperbolic lines are subhyperbolic.

3.3. *Let U be subhyperbolic. Then*

(1) *For all $v, w \in V - U\theta$, Q_v restricted to U is similar to Q_w restricted to U, so the form $Q_U = (Q_v)_{|U}$, $v \in V - U\theta$, is defined up to similarity.*

(2) *A subspace W of U is singular if and only if (W, Q_U) is totally singular.*

(3) *For all $A, B \leq U$, $A \leq B\Delta$ if and only if A is Q_U-orthogonal to B.*

In examples (1)-(4) each subhyperbolic space U is contained in a unique maximal subhyperbolic space $\Phi(U)$.

Define a triple (A_1, A_2, A_3) of points to be ***special*** if for all distinct i, j, k, A_i is singular, $A_i + A_j$ is hyperbolic, and $A_k \not\leq (A_i + A_j)\theta$. The plane $\pi = A_1 + A_2 + A_3$ is a ***special plane***. All special planes over F are isometric to π and contain just three singular points. Observe that $V = \pi \oplus \pi\theta$. Also if S is a set of singular points then $\langle S \rangle$ is brilliant if and only if S contains no special triple.

In examples (1) - (4), $\pi\theta = W_{12} \oplus W_{13} \oplus W_{23}$, where $W_{ij} = (A_i + A_j)\Delta$, and $\Phi(A_i + A_j) = A_i + A_j + W_{ij}$. Further the building for the isometry group of $(V, \mathcal{F})$ can be realized as (essentially) the collection of singular and maximal subhyperbolic subspaces of V with incidence defined by inclusion.

3.4. *Let $G \leq O(V, f)$ be irreducible on V, f nontrivial, and $V = V_1 \oplus\theta \ldots \oplus\theta V_r$ with (V_i, f) indecomposable. Then*

(1) *G permutes $\{V_1, \ldots, V_r\}$ transitively and $N_G(V_i)$ is irreducible on V_i.*

(2) *If F is algebraically closed and G is connected then (V, f) is indecomposable.*

Proof: As G is irreducible on V and f is nontrivial, $V\theta = 0$. If $V = U \oplus\theta W$ and $v \in V$ then $v = u + w$ with $u \in U$, $w \in W$, and $v\Delta = u\Delta^U \oplus w\Delta^W$. Hence if we define $M(U)$ to consist of those $x \in U^\#$ with $x\Delta^U$ maximal, then as $V\theta = 0$, we conclude $M(V) = \bigcup_i M(V_i)$. So as G is irreducible on V, $V_i = \langle M(V_i) \rangle$ and (1) holds. Under the hypothesis of (2), G has no subgroup of finite index, so (1) implies (2).

4. Some Problems

The nature of 3-forms on V will be dependent on our field F, so at least initially let us assume that F is finite or algebraically closed. Even under this assumption, a description of all 3-forms seems out of the question. For example as $dim(S^3(V))$ is greater than $dim(GL(V))$ for $dim(V) > 3$, there are an infinite number of equivalence classes of symmetric trilinear forms on V if $dim(V) > 3$ and F is algebraically closed. So our first problem is to find interesting classes of 3-forms which can be naturally defined and admit strong results. As a group theorist, I think first of forms with big isometry groups. For example:

Problem 1. *Let F be an algebraically closed field. Describe up to isometry all 3-forms $(V, \mathcal{F})$ such that $O(V, \mathcal{F})^\circ$ is irreducible on V.*

To illustrate some techniques available to attack this problem, and the scope of the problem, I sketch a proof of:

4.1. *Assume F is algebraically closed with $p = char(F) \neq 2$, $G \cong SL_2(F)$, and V is a rational irreducible FG-module. Then*

(1) *G preserves a nontrivial trilinear form f on V if and only if $dim(V)$ is odd, in which case f is determined up to a scalar multiple and f is symmetric or alternating for $dim(V) \equiv 1, -1 \bmod 4$, respectively.*

(2) *If $V = M(k_0\lambda_1) \otimes M(k_1\lambda_1)^{\sigma_1} \otimes \cdots \otimes M(k_r\lambda_1)^{\sigma_r}$, where $\sigma_i : a \mapsto a^{p^i}$ is the Frobenius map, then $f = f_1 \otimes \ldots \otimes f_r^{\sigma_r}$, where f_i is the nontrivial G-invariant form on $M(k_i\lambda_1)$.*

(3) *If V is restricted then one of:*
 (a) *$G = O(V, f)^\circ$.*
 (b) *$dim(V) = 3$ and $O(V, f)^\circ = SL(V)$.*
 (c) *$dim(V) = 7$ and $O(V, f) \cong G_2(F)$.*

Proof: Let $X = S$ or Λ and $sgn(X) = +1$ or -1, respectively. Now $V \cong M(k\lambda_1)$ has high weight $k\lambda_1$. If $p = 0$ or $2k < p$ then $V \otimes V$ is the direct sum of distinct irreducibles of restricted high weight, while if $k < p \leq 2k$ then all composition factors of $V \otimes V$ are restricted or $M((2j - p)\lambda_1) \otimes M(\lambda_1)^{\sigma_1}$, where $\sigma_i : a \mapsto a^{p^i}$ is the Frobenius map and $(p + 1)/2 \leq j \leq k$. In either case $Hom_{FG}(V, X^2(V)) \cong F$ if $n = dim(V)$ is odd and $sgn(X) = (-1)^{(n-1)/2}$, while $Hom_{FG}(V, X^2(V)) = 0$ otherwise. Thus (1) holds in this case by 2.1 and 2.2 (and their analogues for alternating forms).

In general by the Steinberg Tensor Product Theorem, either $p = 0$ and V is restricted or V has a unique representation as in 4.1.2. In the latter case as

$$S^2(A \otimes B) \cong S^2(A) \otimes S^2(B) \oplus \Lambda^2(A) \otimes \Lambda^2(B), \text{ and}$$
$$\Lambda^2(A \otimes B) \cong S^2(A) \otimes \Lambda^2(B) \oplus \Lambda^2(A) \otimes S^2(B),$$

we have $X^2(V)$ is the direct sum of terms

$$X_0^2(M(k_0\lambda_1)) \otimes \cdots \otimes X_r^2(M(k_r\lambda_1))^{\sigma_r},$$

where $X_i = S$ or $\wedge$ and $\prod_i sgn(X_i) = sgn(X)$. Appealing to the uniqueness in the Tensor Product Theorem and remarks in the previous paragraph, we conclude that (1) holds again.

Thus (1) holds. Notice (1) imples (2). Finally G is an irreducible primitive subgroup of $Y = O(V, f)^\circ$, so by Seitz [7], either (a) or (c) of (3) holds or $Y = SL(V)$ or $O(V)$. Then by 2.1 and 2.2, $Hom_{FY}(V, \Lambda^2(V^*)) \neq 0$, which forces (3.b) to hold.

Partial results on Problem 1 might help suggest interesting classes of 3-forms. Certain minimal assumptions suggest themselves. For example we should certainly assume $Rad(\mathcal{F}) = V\theta = 0$. Also 3.1 suggests we consider 3-forms $(V, \mathcal{F})$ such that V is generated by singular points. Perhaps we should assume $\mathcal{F}$ is indecomposable in the sense of section 3. The discussion in section 3 suggests a stronger set of assumptions:

Problem 2. *Describe all 3-forms $\mathcal{F}$ such that $Rad(\mathcal{F}) = 0$, V is generated by a set S of singular points, and for all special triples (A_1, A_2, A_3) from S, we have:*

$$(A_1 + A_2 + A_3)\theta = (A_1 + A_2)\Delta \oplus (A_1 + A_3)\Delta \oplus (A_2 + A_3)\Delta.$$

Problem 3. *Describe all 3-forms $\mathcal{F}$ such that $Rad(\mathcal{F}) = 0$, V is generated by a set S of singular points, and for each $\langle x\rangle \in S$, $dim(V/x\Delta)$ is small.*

An example of such a result is the following elementary observation, which is essentially 1.7 in [2]:

4.2. *Assume $\mathcal{F}$ is a 3-form on V, $Rad(\mathcal{F}) = 0$, and S is a set of singular vectors such that $V = \langle S\rangle$ and $dim(V/s\Delta) = 2$ for all $s \in S$. Then $V = \bigoplus \Theta_{P\in\mathcal{P}} P$ with $\mathcal{P}$ a set of special planes such that $S = \bigcup_{P\in\mathcal{P}}(S \cap P)$ and $O(V, \mathcal{F})$ permutes $\mathcal{P}$.*

As I am particularly interested in the subgroup structure of the simple groups and as the subgroup structure of the groups E_6, F_4 and ${}^2E_6$ can be described in terms of the E_6-form, I would be particularly interested in a result strong enough to describe all 3-forms occuring as restrictions of the E_6-form to interesting subspaces of the E_6-module. Such a result might significantly simplify the study of the subgroup structure of these groups.

References:

1. M. Aschbacher, Some multilinear forms with large isometry groups, Geom. Ded. 25 (1988), 417–465.
2. M. Aschbacher, Chevalley groups of type G_2 as the group of a trilinear form, J. Alg. 109 (1987), 193–259.
3. M. Aschbacher, The 27-dimensional module for E_6, I-IV; I, Inven. Math., 89 1987, 159–195; II, J. London Math. Soc. 37 (1988), 275-293; III, Trans. AMS (to appear); IV, J. Alg. (to appear).
4. E. Cartan, "Oeuvres Completes", Gauthier-Villars, Paris, 1952.
5. A. Cohen and A. Helminck, Trilinear alternating forms on a vector space of dimension 7, Comm. Alg., 16 (1988), 1–26.
6. L. Dickson, A class of groups in an arbitrary realm connected with the configuration of the 27 lines on a cubic surface, Quarterly J. Math., 33 (1901), 145–173.
7. G. Seitz, The maximal subgroups of classical algebraic groups, Memoirs AMS 365 (1987), 1–286.

Local recognition of graphs, buildings, and related geometries

Arjeh M. Cohen

A brief overview of results and problems concerning the local recognition of graphs is given, and special attention is drawn toward graphs related to buildings.

1. Introduction

Let Δ be a graph. A graph Γ is said to be *locally* Δ if, for each $\gamma \in \Gamma$, the subgraph (induced on) $\Gamma(\gamma)$ (i.e., the set of vertices adjacent to γ, also called the *neighborhood* or *link* of Γ), is isomorphic to Δ. More generally, when $\mathcal{X}$ is a class of graphs, we say that Γ is *locally* $\mathcal{X}$ if, for each $\gamma \in \Gamma$ (read γ a vertex of Γ) the subgraph $\Gamma(\gamma)$ is isomorphic to a member of $\mathcal{X}$. Of course, by saying that a graph is *locally grid* (a *grid* being a graph that can be written as a product of two cliques), we mean that it is locally $\mathcal{X}$, where $\mathcal{X}$ stands for the class of all grids.

Although it was probably apparent to quite a few finite group theorists that some of the local characterization problems in the classification of finite simple groups can be phrased in terms of graphs, Hall and Shult (to the best of my knowledge) were the first to strip the group theory so as to obtain intriguing problems of merely graph-theoretic nature: given an 'interesting' class $\mathcal{X}$, determine all graphs which are locally $\mathcal{X}$. We shall refer to it as the *local recognition problem (with respect to $\mathcal{X}$).* The theory of buildings also led to local characterization questions for geometries, which can be reduced to local recognition problems for graphs. These are the two main reasons why I want to survey what is known on the local recognition problem.

Much of the material in this paper can be found scattered in [BCN 1989]. For notions of shadow space, point-line geometry and the like, the reader is referred to [Coh 1986].

The first major paper dealing with the topic known to me is [BuHu 1977] treating graphs that are locally generalized quadrangles. For quite a different angle, see [BHM 1980], where the question is raised whether, for Δ a specified tree, a graph Γ exists that is locally Δ; see also [Hall 1985] for a sequel.

For any graph Γ, denote by $\mathcal{X}(\Gamma)$ the set of all graphs of the form $\Gamma(\gamma)$ for $\gamma \in \Gamma$. Clearly Γ need not be the only graph (up to isomorphism) that is locally $\mathcal{X}(\Gamma)$. For instance, the disjoint union of two copies of Γ has the same property. To remove this obstacle it suffices to demand that Γ be connected. Another instance: any polygon is locally $2K_1$, the disjoint union of two vertices. But then the infinite path is a cover of them all, so constitutes a universal solution. (More generally, if $d(\cdot,\cdot)$ denotes the usual graph-theoretic distance in Γ and A is a subgroup of $Aut\Gamma$ satisfying $d(\gamma,\gamma^a) \geq 4$ for all $\gamma \in \Gamma$ and $a \in A$, then the natural quotient graph Γ/A is locally $\mathcal{X}(\Gamma)$.) A third instance: there are precisely two graphs (up to isomorphism) that are locally a cube (see below); they do not seem to have much more than this property in common. Thus, the best outcome one may expect of local recognition with respect to a class $\mathcal{X}$ is a list of well-described graphs such that any connected graph that is locally $\mathcal{X}$ is a suitable quotient of one of the graphs in the list.

Now, take into consideration a 'natural' class of graphs, such as the dual polar spaces. Their local structure is rather uninformative (a disjoint union of cliques) and does not suffice to pin them down. A partial solution is to extend the notion of local recognition to bigger neighborhoods: for any natural number h, the h-neighborhood of $\gamma \in \Gamma$, notation $\Gamma_{\leq h}(\gamma)$, is the subgraph induced on the set of all vertices at distance at most h to γ. Thus the 1-neighborhood of γ, sometimes denoted by $\gamma^\perp$ rather than $\Gamma_{\leq 1}(\gamma)$, is the complete join of the vertex γ and $\Gamma(\gamma)$. We say that a class Ψ of connected graphs is *locally h-recognizable* if every connected graph Δ with the property that for each $\delta \in \Delta$ there are $\Gamma \in \Psi$ and $\gamma \in \Gamma$ such that $\Delta_{\leq h}(\delta) \cong \Gamma_{\leq h}(\gamma)$ is a suitable quotient of a member of Ψ. In [Coh 1986] the following question was raised: given a 'natural' class of graphs, what is the minimal h such that it is locally h-recognizable? For the class of all dual polar graphs, $1 < h \leq 8$. In general h need not exist. If Ψ consists of a single finite member Γ that is regular (i.e., all vertices have the same finite valency), then h exists.

The minimal h for the point graph Γ of a geometry is obviously related to axiom systems characterizing that geometry: any axiom system should involve a condition that cannot be checked in $\Gamma_{\leq j}(\gamma)$ for $\gamma \in \Gamma$ and $j < h$; in other words, if there is a number j and an axiom system such that the validity of each of its axioms can be verified by mere knowledge of the structure of $\Gamma_{\leq j}(\gamma)$ for $\gamma \in \Gamma$, then $h \leq j$.

2. Some examples and elementary observations

Often, if Δ has diameter at most 2 and there are plenty of edges, the locally Δ graphs can be found. The most extreme case:

Proposition: *If $\Delta = K_n$ (the complete graph, or clique, of size n), then K_{n+1} is the only connected graph (up to isomorphism) that is locally Δ.*

There need not always be a graph that is locally Δ. Take $\Delta = K_{1,2}$, where $K_{m_1,\ldots,m_t}$ is the complete t-partite graph with parts of size $m_1, \ldots, m_t$. More generally,

Proposition (cf. [BCN 1989]): *Let Γ be a connected graph that is locally complete multipartite. Then Γ is either triangle-free or complete multipartite. In particular, if there are $(m_i)_{1\leq i\leq t}$ and $\gamma \in \Gamma$ such that $\Gamma(\gamma)$ is locally $K_{m_1,\ldots,m_t}$, then $m_1 = \ldots = m_t$, and Γ is the complete $t+1$-partite graph with parts of size m_1.*

Existence may also depend on finiteness:

Proposition ([BHM 1980]): *Let Δ be the tree on 8 vertices with 3 vertices of valency 3 and 5 of valency 1. There exists an infinite graph that is locally Δ; but no finite one.*

The existence proof is based on a 'free construction', the non-existence proof on a counting argument (the number of edges on a unique triangle can be counted in two ways that lead to distinct answers).

The most 'classical' example of local recognition of graphs inspired by groups is related to $P\Sigma L(2,25)$. This group appears as a sporadic example in the odd transposition papers of [Asch 1973]. The transpositions in question are the 65 outer involutions of a single class D. Let Γ be the graph whose vertex set is D and in which two distinct vertices are adjacent whenever they commute (the so-called 'commuting involutions' graph on D). Then $\Gamma(\gamma)$ can be identified with the commuting involutions graph on the class of transpositions in Sym_5, the symmetric group on 5 letters. In other words, Γ is locally Petersen. The characterization of all groups G having a class of transpositions D such that, for $d \in D$, the set $X = D \cap C_G(d)$ maps to the set of transpositions in Sym_5 under the composition of morphisms $C_G(d) \to C_G(d)/C_G(X) \cong Sym_5$ is a corollary of the determination of all graphs that are locally Petersen:

Proposition ([Hall 1980]): *Apart from the above graph related to $P\Sigma L(2,25)$, there are two connected locally Petersen graphs, viz. the complement of the Johnson graph $J(7,2)$ and a 3-cover of this graph (with automorphism group $3 \cdot Sym_7$).*

The proof of this result is elementary, but requires work. The main step is the determination of the structure induced on the set of points adjacent to two points at distance 2.

Local characterizations of Lie graphs (gotten as point graphs of shadow spaces of buildings) are best illustrated by the graphs that are locally polygons. Recall (cf. [BCN 1989]) that the Coxeter graph of type $M_{n,i}$ is the graph obtained from the Coxeter group W of type $M = (m_{jk})_{1\leq j,k\leq n}$ in the following way: set $r = r_i$; vertices are the cosets with respect to the subgroup $V = \langle r_j \mid j \neq i \rangle$; two distinct vertices xV, yV are adjacent if $y^{-1}x \in VrV$. Let $M = (m_{ij})_{1 \leq i,j \leq 3}$, where $m_{12} = 3$, $m_{13} = 2$, and $m_{23} = m$. The Coxeter graph $\Gamma^{(m)} = M_{3,1}$ is a locally m-gon graph. It is finite if and only if $m \leq 5$. [This graph is nothing but the 1-shadow of the thin building (apartment) of type M; edges correspond to objects of type 2, triangles to objects of type 3.]

Proposition: *Let $m \geq 2$. If Γ is a connected graph that is locally the m-gon, then there is a group A of automorphisms of $\Gamma^{(m)}$ such that $\Gamma \cong \Gamma^{(m)}/A$.*

The proof consists of constructing a chamber system of type M from Γ with the above notions for objects, and letting W operate on it as a set of permutations. This gives a morphism $\Gamma^{(m)} \rightarrow \Gamma$. The group A consists of all elements in $Aut\ \Gamma^{(m)} \cong W$ fixing the fiber of a given chamber of Γ.

The following elementary and charming argument shows that there is some control over graphs that are locally collinearity graphs of finite polar spaces. Here, we define a *Zara graph* to be a finite non-complete graph with the properties that all maximal cliques have the same size, that the number $e := \mid \gamma^{\perp} \cap M \mid$ is constant for all vertices γ and maximal cliques M not containing γ, and that, for γ and M as before, there exists a maximal clique on γ meeting M in less than e points. (Thus, $e > 0$.)

Lemma ([Pasi 1989]): *Let Γ be a connected locally Zara graph. Then all maximal cliques of Γ have the same size, m say, and the diameter of Γ is bounded from above by $m - 1$.*

Proof: The statement about constant maximal clique size follows directly from the corresponding Zara graph property and connectedness of Γ. Suppose $\gamma \in \Gamma$ and M is a maximal clique. The *distance from γ to M*, denoted by $d(\gamma, M)$, is the minimum over all distances $d(\gamma, \delta)$ for $\delta \in M$. We assert that $d(\gamma, M) = i$ implies $i \leq \mid \Gamma_i(\gamma) \cap M \mid -1$. It clearly proves the lemma.

The assertion follows from induction on i: if $i = 0$, then the inequality reads $0 \leq 0$. So let $i > 0$, and take $\delta \in \Gamma_i(\gamma) \cap M$, $\epsilon \in \Gamma_{i-1}(\gamma) \cap \Gamma(\delta)$. Pick a maximal clique M' on $\{\epsilon, \delta\}$ such that $\mid M \cap M' \mid < \mid \epsilon^{\perp} \cap M \mid$. Then $d(\gamma, M') = i - 1$, so the induction hypothesis yields that $\Gamma_{i-1}(\gamma) \cap M'$ contains at least i vertices. Counting edges between $M \setminus M'$ and $M' \setminus M$ we find that there are at least i vertices in $\bigcup_{\zeta \in \Gamma_{i-1}(\gamma) \cap M'} \zeta^{\perp} \cap M \setminus M'$, whence at least $\mid M \cap M' \mid + i$ vertices in $\Gamma_{\leq i}(\gamma) \cap M$, proving the assertion as $\delta \in M \cap M'$. QED

In many cases, the diameter bound can be considerably improved (although in the locally generalized quadrangle case of the above lemma, it is best possible!); here we have merely derived that there exists a bound, implying that the solutions to the local recognition problem will only involve finite graphs.

3. Covers and imprimitivity

In group theory, the notion of imprimitivity helps to reduce arbitrary permutation representations to primitive ones. For association schemes, there is a satisfactory counterpart, cf. [BCN 1989]. Graph-theoretically, there is no unified approach. In [Asch 1976], a partial interpretation has been given in terms of *contractions*, i.e., surjective morphisms of graphs $\Gamma \to \Delta$ such that any two distinct vertices δ and δ' are adjacent if and only if every vertex in the fiber of δ is adjacent to every vertex in the fiber of δ'. One reason why it is partial is that contraction could also be considered for the graph $\Gamma_{\mathcal{I}}$, whose vertex set coincides with that of Γ and in which two vertices are adjacent whenever they are at distance i in Γ for some $i \in \mathcal{I}$.

For example, by the contraction criterion, the disjoint union of cliques is imprimitive. If Γ is the 3-cover in the locally Petersen proposition, the graph Γ_3 is a disjoint union of cliques; the corresponding primitive quotient – obtained by calling two cliques X, Y of Γ_3 adjacent if there is an edge of Γ joining a vertex from X with one from Y – is the Johnson graph $J(7,2)$. The surjective morphism defined on Γ thus obtained is an isomorphism when restricted to $\Gamma_{\leq 1}(\gamma)$ for each $\gamma \in \Gamma$.

We give two examples regarding the question of finding local conditions that are sufficient for Γ to be imprimitive.

Proposition ([JoSh 1988]): *Let Γ be a graph. Call γ and δ equivalent whenever $\gamma^\perp = \delta^\perp$. Then $\Gamma \to \Gamma/\equiv$ is a contraction. Moreover, $\Gamma/\equiv$ is reduced (i.e., $\gamma^\perp = \delta^\perp$ implies $\gamma = \delta$). If Γ is finite, regular, and reduced, then Γ is locally nondegenerate (a graph is called* nondegenerate *if no vertex is adjacent to all others).*

Proposition (cf. [BCN 1989]): *Let Δ be a finite graph whose complement is not connected, and suppose Γ is a connected graph that is locally Δ. If $\Gamma(\gamma) \cap \Gamma(\delta)$ has the same size for all $\gamma, \delta \in \Gamma$ at distance 2, then Γ is a complete multipartite graph.*

4. Coxeter graphs

Since the Coxeter graphs $M_{n,i}$ are shadow spaces of apartments in buildings, local recognition of these graphs serves as a prelude to local recognition of the shadow spaces of buildings themselves. If M is irreducible and spherical, we shall adopt Bourbaki's node labeling of the diagram for M (cf. [Bourb 1968]). The graph $A_{n,i}$ is the Johnson graph $J(n+1, i)$. It is locally an $i \times (n-i)$ grid. The following result is known in various guises and special cases (see [BCN 1989] for further details and references).

Theorem: *Let Γ be a connected locally grid graph. Suppose that no connected component of the common neighbor subgraph $\Gamma(\gamma) \cap \Gamma(\delta)$ of two vertices at distance 2 has strictly more than 4 points. Then Γ is a Johnson graph or the quotient of a Johnson graph $J(2n, n)$ by a suitable involutory automorphism.*

Connected locally $2 \times n$ grid graphs necessarily satisfy the supposition. (Locally $J(k, 2)$ graphs have been studied in [Neum 1985].)

There are precisely two locally 3×3 grid graphs, namely the Johnson graph $J(6, 3)$ and the complement of the 4×4 grid. More generally, [Hall 1989] has identified these graphs with line graphs of certain Fischer spaces. As a consequence all locally $3 \times n$ grid graphs are known. [BlBr 1984] proved that, besides $J(8, 4)$ and the (unique) quotient by the automorphism 'taking complements', there are precisely two more connected locally 4×4 grid graphs, each having 40 vertices and diameter 3.

[Hall 1987] studied local recognition of the maximal distance graphs $J(n, i)_i$ of $J(n, i)$. He showed that if $n \geq 3i + 1$ any connected graph that is locally $J(n, i)_i$ must be isomorphic to $J(n + i, i)_i$. The bound is sharp in the sense that for $(n, i) = (6, 2)$ (cf. [BuHu 1977]), there are two more connected graphs that are locally $J(6, 2)_2$, viz. the complements of an elliptic and of a hyperbolic quadric in $Sp(4, 2)$.

The graphs $D_{n,1}$ and $B_{n,1} = C_{n,1}$ are all isomorphic to the n-partite graph with parts of size 2, so are dealt with above. At the other end node of the Coxeter diagram we have the n-cube $B_{n,n}$ and its halved graph $D_{n,n}$. Now the hypercubes are clearly not locally 1-recognizable. On the other hand, they are 3-recognizable as Brouwer worked out, cf. [BCN 1989]. Finally, the halved n-cubes are locally $J(n, 2)$. Such graphs are treated by [Neum 1985].

The graph $E_{6,1}$, known as the *Schläfli graph*, is the complement of the collinearity graph of the unique generalized quadrangle of order $(2, 4)$; it is the unique connected graph that is locally $D_{5,5}$. The graph $E_{7,7}$, known as the *Gosset graph*, is the unique connected locally Schläfli graph. The graph $E_{8,8}$ is the unique connected graph that is locally $E_{7,7}$.

The graph $F_{4,1}$ is locally a cube. The only other connected locally cube graph is the complement of the 4×5 grid (see [Bus 1983]).

Concerning $H_{n,1}$, the locally icosahedral graphs (so $n = 4$) can be determined in the same way as the locally pentagon graphs (so $n = 3$), see [BBBC 1985]. (Two references on the nonspherical case of locally dodecahedron graphs are: [Cox 1954] and [VC 1985].)

5. Point-line geometries of spherical buildings

During the last few years, the thick analogues of the graphs $M_{n,i}$ for M of spherical type, i.e., the collinearity graphs of shadow spaces of spherical buildings – also called *Lie graphs* – have received some more attention beyond what has been described in [Coh 1986]. Here we content ourselves with an update of those notes. See [BCN 1989] for several properties of Lie graphs. Using (a slight variation of) the concept of parapolar space of polar rank 3, [HaTh 1988] have given an axiom system characterizing the finite Lie graphs of type $A_{n,i}$ ($2 \leq i \leq n-1$), $C_{n,n-2}$ ($n \geq 3$), $D_{n,n-1}$ ($n \geq 4$), $E_{7,4}$, $E_{8,5}$, and $F_{4,1}$. The axioms beyond those for a parapolar space only concern configurations of symplecta intersecting in a line (which in fact follow from 1-local knowledge), so the result provides a local 2-recognition of these graphs. See [Shult 1989] for another local characterization of some of these graphs. The Grassmannians of polar spaces have also been characterized: Starting with a parapolar space that is locally a graph whose reduced graph is a Lie graph of type $A_m \cup B_n$ ($m \geq 1$ and $n \geq 3$) [Hanss 1987] constructed globally defined subspaces of type $A_{n,2}$; from this he derived (with existing methods) that such a parapolar space is a suitable quotient of a Lie graph of type $B_{m+n+1,m+1}$. Thus, the Grassmannians of polar spaces are locally 2-recognizable. [Shult 1988] has shown that 4-shadow spaces of buildings with affine diagram F_4 (so the diagram has rank 5 diagram and maximal singular subspaces have rank 2) are locally 2-recognizable.

6. Beyond buildings

Now that the shadow spaces of spherical buildings are reasonably well understood, it is time to explore related geometries and graphs that are no longer of Lie type.

Removing a geometric hyperplane H (i.e., a subset such that each line either has a single point or all of its points inside of it) from a shadow space S of a spherical building leads to a partial linear space on $S \setminus H$ with the same local structure as S (that is, for $x \in S \setminus H$, the lines and planes of $S \setminus H$ on x form a space isomorphic to the space of lines and planes of S on x). We call such a space an *affine space* because if S is a projective space, the usual affine space appears, and (hence) if S is arbitrary, every (singular) plane of S not contained in H becomes a 'classical' affine plane in the space $S \setminus H$. In [CoSh 1987], a local recognition theorem is given for affine polar spaces of rank ≥ 3. Thus the classification of affine polar spaces amounts to that of classifying geometric hyperplanes of polar spaces. If the shadow space S is embedded in a projective space, geometric hyperplanes can be obtained as intersections of projective hyperplanes of the ambient projective space not containing all points of the embedded geometry. In [*loc. cit.*] it is also shown that, in case S is a polar space of rank ≥ 3, every hyperplane can be obtained by this construction if a projective embedding of S exists, and is of the form $x^\perp$ for some point x of S otherwise. In general, that is, for an arbitrary shadow space S, the local recognition of corresponding affine spaces is complicated by the fact that the geometric hyperplanes come in many kinds (there are at least as many orbits under the corresponding group G of Lie type as the number of projective hyperplane orbits of a projective space in which S is embedded, which equals the number of point orbits of the corresponding G-module).

The two sources of inspiration for local recognition mentioned at the outset both yield classes of geometries that merit further study. First, as pointed out by [Lyons 1988], the classification of finite simple groups uses 'component type' theorems which are related to commuting tori graphs (i.e., graphs whose vertex sets are conjugacy classes of 1-dimensional tori or semi-simple elements of a group of Lie type, and in which two vertices are adjacent when they commute). Second, according to a result of [Seitz 1974] the primitive parabolic permutation representations (i.e., those on the point sets of our Lie graphs) are the only primitive permutation representations of the corresponding groups of Lie type whose permutation ranks do not depend on the order of the underlying field. One step beyond these are the transitive permutation representations of groups of a given Lie type that have a permutation rank that is linear as a function of the order of the underlying field. In various cases, an association scheme can be found whose number of classes is independent of the order of the field.

An example of what I have in mind is the permutation representation of $A_n(q) \cong PSL(n+1,q)$ on the set P of antiflags (= nonincident point hyperplane pairs) in the associated projective space. The permutation rank of the full group $AutA_n(q)$ (graph automorphisms included) on P is $q+3$ (cf. [Dar 1986]). Nevertheless, the projective geometry naturally produces an association scheme of 6 classes (one class is empty if the field has order 2). Taking two antiflags to be adjacent when the point of each of them lies in the hyperplane of the other, we turn P into the commuting tori graph described above with respect to the class of tori whose centralizer has 'simple component' of type A_{n-1}. (It is not always true that small permutation rank, say linear in the order q of the underlying field, implies that the point stabilizer normalizes a torus, the case of ${}^2B_2(q)$ in $Sp(4,q)$ being a counterexample).

Other examples come from association schemes on the nonisotropic points in an orthogonal or unitary space. The beautiful result [HaSh 1985] on locally cotriangular graphs is the first local recognition theorem for graphs related to those schemes (only very small fields appear in the conclusion of the theorem). The characterization of affine polar spaces in [CoSh 1987] uses the same ideas of proof. This is not so surprising as it may seem at first sight because certain spaces, whose points are the nonisotropic points, occur as quotients of suitable affine polar spaces. For example, consider the unitary polar space S embedded in a projective space over $\mathbb{F}_{q^2}$, and the geometric hyperplane $H = x^{\perp} \cap S$ of all points of S perpendicular (with respect to the unitary form associated with S) to some nonisotropic point x. The relation 'having distance 3 in the collinearity graph' is an equivalence relation on $S \setminus H$. The quotient space of $S \setminus H$ by the equivalence relation is the space N induced on the set of all nonisotropic points of $x^{\perp}$ whose lines are the tangents to H. The planes are affine again, but given a point x and a line l not on x, there are either 0, q, or q^2 points of l collinear with x. Recently, H. Cuypers has further extended the methods of [CoSh 1987] and [Pasi 1988] to obtain a characterization of the space N, valid for all finite fields.

References:

[Asch 1973] M. Aschbacher, *On finite groups generated by odd transpositions, II, III, IV*, J. Algebra, **26**(1973), 451–459, 460–478, 479–491.

[Asch 1976] M. Aschbacher, *A homomorphism theorem for finite graphs*, Proc. Amer. Math. Soc., **54**(1976), 468–470.

[BHM 1980] A. Blass, F. Harary & Z. Miller, *Which trees are link graphs?*, J. Combinatorial Th. (B), **29**(1980), 277–292.

[BlBr 1989] A. Blokhuis & A.E. Brouwer, *Locally 4-by-4 grid graphs*, J. Graph Th., **13**(1989), 229–244.

[BBBC 1985] A. Blokhuis, A.E. Brouwer, D. Buset & A.M. Cohen, *The locally icosahedral graphs*, pp. 19–22 in: Finite geometries, Proc. Winnipeg 1984 (eds.: C.A. Baker & L.M. Batten), Marcel Dekker, New York, Lecture Notes in Pure and Applied Math. 103, 1985.

[Bourb 1968] N. Bourbaki, *Groupes et algèbres de Lie, Chap. 4,5 et 6*, Hermann, Paris, 1968.

[BCN 1989] A.E. Brouwer, A.M. Cohen & A. Neumaier, *Distance-Regular Graphs*, Ergebnisse der Math. Vol. 18, Springer Verlag, Berlin, 1989.

[BuHu 1977] F. Buekenhout & X. Hubaut, *Locally polar spaces and related rank 3 groups*, J. Algebra, **45**(1977), 155–170.

[Bus 1983] D. Buset, *Graphs which are locally a cube*, Discrete Math., **46**(1983), 221–226.

[Coh 1986] A.M. Cohen, *Point-line characterizations of buildings*, pp. 191–206 in: Buildings and the geometry of diagrams, Proc. Como 1984 (ed.: L.A. Rosati), Springer Verlag, Berlin, Lecture Notes in Math. 1181, 1986.

[CoSh 1987] A.M. Cohen & E.E. Shult, *Affine polar spaces*, preprint, (1987), 19 pp.

[Cox 1957] H.S.M. Coxeter, *Regular honeycombs in hyperbolic space*, pp. 155–169 in: Proc. Intern. Congress Math., Amsterdam 1954 (eds.: J.C.H. Gerretsen & J. de Groot), Groningen, 1957.

[Dar 1986] M.R. Darafsheh, *On a certain permutation character of the General Linear Group*, Comm. Algebra, **14**(1986), 1343–1355.

[Hall 1980] J.I. Hall, *Locally Petersen graphs*, J. Graph Th., **4**(1980), 173–187.

[Hall 1985] J.I. Hall, *Graphs with constant link and small degree or order*, J. Graph Th., **8**(1985), 419–444.

[Hall 1987] J.I. Hall, *A local characterization of the Johnson scheme*, Combinatorica, **7**(1987), 77–85.

[Hall 1989] J.I. Hall, *Graphs, geometry, 3-transpositions and symplectic* $\mathbb{F}_2$*-transvection groups*, Proc. London Math. Soc., **58**(1989), 89–111.

[HaSh 1985] J.I. Hall & E. Shult, *Locally cotriangular graphs*, Geom. Dedicata, **18**(1985), 113–159.

[Hanss 1987] G. Hanssens, *A technique in the classification theory of point-line geometries*, Geom. Dedicata, **24**(1987), 85–101.

[HaTh 1988] G. Hanssens & J. Thas, *Pseudopolar spaces of polar rank three*, Geom. Dedicata, **22**(1988), 117–135.

[JoSh 1988] P. Johnson & E.E. Shult, *Local characterizations of polar spaces*, Geom. Dedicata, **28**(1988), 127–151.

[Lyons 1988] R. Lyons, *lecture given at Oberwolfach*, Gruppen und Geometrien, 1988.

[Neum 1985] A. Neumaier, *Characterization of a class of distance regular graphs*, J. reine angew. Math., **357**(1985), 182–192.

[Pasi 1988] A. Pasini, *On locally polar geometries whose planes are affine*, preprint, (1988), Siena, Italy.

[Pasi 1989] A. Pasini, *On the diameter of extended generalized quadrangles*, preprint, (1989), Siena, Italy.

[Seitz 1974] G.M. Seitz, *Small rank permutation representations of finite Chevalley groups*, J. Algebra, **28**(1974), 508–517.

[Shult 1989] E. Shult, *Characterizations of spaces related to metasymplectic spaces*, Geom. Dedicata, **30**(1989), 325–371.

[Shult 1989] E. Shult, *A remark on Grassmann spaces and half-spin geometries*, preprint, (1989), Manhattan, Kansas.

[VC 1985] P. Vanden Cruyce, *A finite graph which is locally a dodecahedron*, Discrete Math., **54**(1985), 343–346.

Generalized Polygons and s-Transitive Graphs

Richard Weiss*

1. Introduction

Let Γ be a connected, undirected graph and G a subgroup of aut(Γ). For each vertex x, we will denote by $\Gamma(x)$ the set of neighbors of x in Γ. An s-path (for any $s \geq 0$) is a sequence $(x_0, x_1, \ldots, x_s)$ of $s+1$ vertices x_i such that $x_i \in \Gamma(x_{i-1})$ for $1 \leq i \leq s$ and $x_i \neq x_{i-2}$ for $2 \leq i \leq s$. The graph Γ is called *locally s-transitive* (with respect to G) if for each vertex x, the stabilizer G_x acts transitively on the set of s-paths beginning at x. If, in addition, G acts transitively on the vertex set of Γ, and hence on the set of all s-paths in Γ, then Γ is called *s-transitive* with respect to G.

The girth of a graph Γ is the length of a shortest circuit in Γ (assuming circuits exist). The diameter of Γ is the maximal distance between two points of Γ, which may be infinite. A graph Γ is called a *generalized n-gon* (for some $n \geq 2$) if it is a bipartite graph of diameter n and girth $2n$. A generalized n-gon Γ with $|\Gamma(x)| = 2$ for each vertex x is just the incidence graph of an ordinary n-gon. A generalized 2-gon is a complete bipartite graph. The graph Γ is a generalized 3-gon with $|\Gamma(x)| \geq 3$ for each x if and only if it is the incidence graph of a projective plane.

Note that if Γ is a circuit (i.e. an ordinary n-gon) or a regular tree (i.e. a tree for which $|\Gamma(x)|$ is independent of x) and $G =$ aut(Γ), then Γ is s-transitive with respect to G for all s. There is, however, a simple connection between girth and s-transitivity **[21]**:

(1.1) Proposition: *If Γ is locally s-transitive and $|\Gamma(x)| \geq 3$ for some x, then $s \leq (g+2)/2$, where g is the girth of Γ. Equality holds if and only if Γ is a generalized $(s-1)$-gon.*

The notion of a generalized n-gon is due to J. Tits **[16]**, for whom these graphs were the rank 2 case of his geometries called buildings. Generalized n-gons which are also $(n+1)$-transitive were called cages by W. T. Tutte. It was Tutte who introduced the concept of s-transitivity and who first made the observation (1.1) (for s-transitive rather than locally s-transitive

* Research partially supported by NSF Grant DMS-87000838.

graphs). In [**19-20**], he classified trivalent cages. In the course of this work, he proved the following remarkable theorem:

(1.2) Theorem: *If* Γ *is finite, trivalent and* s*-transitive, then* $s \leq 5$.

We will give here a brief survey of various efforts to extend and apply this and some related (but historically independent) results of Tits. As these developments include the entire theory of amalgams opened up by D. Goldschmidt's seminal paper [**8**] on locally s-transitive trivalent graphs, our choice of results will, of necessity, be very selective. In general, we will describe results which are of a more graph theoretical nature. The reader is referred to [**22**] for an earlier report on s-transitive graphs.

2. Generalized polygons

As mentioned above, a generalized polygon is the rank 2 case of Tits' concept of a building of spherical type. Buildings of spherical type and rank ≥ 3 were classified in [**16**]; for rank $= 2$, such a classification is not feasible since, in particular, this would include a classification of all projective planes.

A building of spherical type and rank ≥ 3 is essentially equivalent to the notion of a group with a BN-pair (see [**16**] or p.78 of [**4**]). If G is a group with a BN-pair of rank two, then there is a generalized n-gon for some n on which G acts locally $(n+1)$-transitively with B the stabilizer of an edge and N the setwise stabilizer of a $2n$-circuit through this edge. In fact:

(2.1) Proposition: *Let* Γ *be a generalized* n*-gon,* Δ *a* $2n$*-circuit in* Γ *and* $\{x, y\}$ *an edge of* Γ. *Let* $\mathrm{aut}^\circ(\Gamma)$ *be the group of type-preserving automorphisms of* Γ *(i.e. those automorphisms which map each vertex to one at even distance) and let* G *be a subgroup of* $\mathrm{aut}^\circ(\Gamma)$. *Let* $B = G_{xy}$ *and let* N *be the setwise stabilizer of* Δ *in* G. *Then the subgroups* B *and* N *form a* BN*-pair of rank* 2 *for* G *if and only if* Γ *is locally* $(n+1)$*-transitive with respect to* G.

Thus, the notion of local s-transitivity is implicit in the notion of a BN-pair of rank 2. To extend the classification of buildings of spherical type to the rank 2 case, a still stronger notion (due to Tits) is required:

(2.2) Definition: *A generalized* n*-gon* Γ *(for* $n \geq 3$*) is called Moufang if for each* $(n-1)$*-path* $(x_1, \ldots, x_n)$*, the subgroup* $G^{[1]}_{x_1,\ldots,x_{n-1}}$ *acts transitively on* $\Gamma(x_n)/\{x_{n-1}\}$.

Here $G^{[1]}_{x_1,\ldots,x_{n-1}}$ denotes the largest subgroup of the stabilizer $G_{x_1,\ldots,x_{n-1}}$ acting trivially on $\Gamma(x_i)$ for $1 \leq i \leq n-1$. Note that the Moufang property implies local s-transitivity for $s = n+1$. In [**17**] (see also [**25**]), Tits showed that Moufang polygons with $|\Gamma(x)| \geq 3$ for each x exist only for $n = 3$, 4, 6 and 8; he then classified Moufang polygons for each of these values of n. The classification for $n = 3$ is a classical result. The case $n = 8$ is done in [**18**], and the cases $n = 4$ and $n = 6$ remain unpublished. If Γ is assumed to be finite, the Moufang condition implies (for instance, by [**23**]) that $B = O_p(B) \cdot H$ for some prime p, where H is the pointwise stabilizer of Δ in G. Generalized polygons with this property were classified in [**6**].

3. The local structure of s-transitive graphs

Let Γ be an arbitrary graph which is s-transitive graph with respect to a group G such that $|G_x| < \infty$ for each vertex x. It was shown in [**26**] that if $s \geq 4$, then

$$PGL_2(q) \leq G_x^{\Gamma(x)} \leq P\Gamma L_2(q) \qquad (*)$$

(up to isomorphism) for each x and some prime power $q = p^m$, where $|\Gamma(x)| = q + 1$. Here $G_x^{\Gamma(x)}$ denotes the permutation group induced by G_x on $\Gamma(x)$. Let $k = q + 1$. The essential ingredients in [**26**] are the result of Thompson [**15**] (see also [**7**], [**23**] and [**36**]), which says that $G_{xy}^{[1]}$ is a p-group for some prime p and each edge $\{x, y\}$, and the p-factorization method of Thompson. The classification of 2-transitive permutation groups of degree k is also required, so the proof of this result depends on the classification of finite simple groups.

Suppose now that $s \geq 4$, so $(*)$ holds. Let $\{x, y\}$ be an edge of Γ. In [**24**], it was shown how to construct a Moufang generalized $(s-1)$-gon $\tilde{\Gamma}$ on which the stabilizers G_x and $G_{\{x,y\}}$ act faithfully as point- and edge-stabilizers. With the classification of Moufang polygons, we reach the following conclusion:

(3.1) Theorem: *If Γ is an s-transitive graph with respect to a group G such that $s \geq 4$ and $|G_x| < \infty$, then $(*)$ holds and either $s = 4$ with q arbitrary or $s = 5$ and $p = 2$ or $s = 7$ and $p = 3$. Moreover, the structure of the subgroups G_x and G_{xy} is determined by their action on the generalized $(s-1)$-gon $\tilde{\Gamma}$.*

A similar result is known for graphs which are only locally s-transitive. Dealing with this case is a much more difficult problem. The famous paper [**8**] of Goldschmidt on the trivalent case showed how to proceed (see [**33**]). Locally s-transitive graphs are generally studied under the heading of amalgams. The amalgam referred to is simply the triple of subgroups $(G_x, G_y; G_{xy})$, where $\{x, y\}$ is an edge of Γ. It was Goldschmidt who introduced the useful technique of replacing Γ by a tree on which the free amalgamated product $G_x *_{G_{xy}} G_y$ acts. (We mention this to emphasize the purely local nature of these results.)

In [**4**], the following major result was proved:

(3.2) Theorem: *Suppose Γ is locally s-transitive for some $s \geq 2$ with respect to a group G such that $|G_x| < \infty$ for each vertex x. Suppose, too, that for some prime p, $G_x^{\Gamma(x)}$ contains as a normal subgroup (up to isomorphism) a group of Lie type of rank 1 defined in characteristic p acting in its standard 2-transitive representation, where the group of Lie type (but not p) may depend on x. Then either $s \geq 4$ and the amalgam $(G_x, G_y; G_{xy})$ is determined by its action on a certain Moufang generalized $(s-1)$-gon $\tilde{\Gamma}$ or $s \leq 3$ and $(G_x, G_y; G_{xy})$ is trivial in an appropriate sense.*

An important open problem is to make this result into a generalization of (3.1) by showing that if $s \geq 4$, then the condition on the permutation groups $G_x^{\Gamma(x)}$ of (3.2) must hold (compare [**26**]).

4. Some applications

Again let Γ be a graph which is s-transitive with respect to a group G such that $|G_x| < \infty$ for each vertex x and suppose that $s \geq 4$, so that (3.1) holds. Recall that $s \leq (g+2)/2$ by (1.1). If we suppose as well that Γ is distance-transitive with respect to G, i.e. that for each i, the group G acts transitively on the set of order pairs (u, v) of vertices such that the distance between u and v is i, then it is easy to see that $s \geq (g-2)/2$. This observation suggests the question whether it is possible to use (3.1) to classify distance-transitive graphs with $s \geq 4$. In fact, distance-transitivity turns out to be unnecessary [**29-32**]:

(4.1) Theorem: *Let Γ be a graph which is s-transitive with respect to a group G such that $|G_x| < \infty$ and $k = |\Gamma(x)| \geq 3$ for each vertex x. Suppose, too, that $s \geq 4$ and $s \geq (g-2)/2$. Let α be the number of vertices of Γ. Then either $s = (g+2)/2$ and Γ is a Moufang generalized polygon or one of the following holds:*

(i) $k = 3$, $s = 4$, $g = 8$, $\alpha = 30$ *and* $G' \cong A_6$,
(ii) $k = 3$, $s = 4$, $g = 9$, $\alpha = 102$ *and* $G \cong L_2(17)$,
(iii) $k = 3$, $s = 4$ *or* 5, $g = 10$, $\alpha = 90$ *and* $G' \cong 3 \cdot A_6$,
(iv) $k = 3$, $s = 5$, $g = 12$, $\alpha = 243$ *and* $G \cong \mathrm{aut}(L_3(3))$,
(v) $k = 3$, $s = 5$, $g = 12$, $\alpha = 486$ *and* $G \cong \mathrm{aut}(GL_3(3))$,
(vi) $k = 3$, $s = 5$, $g = 12$, $\alpha = 650$ *and* $G \cong \mathrm{aut}(L_2(25))$,
(vii) $k = 4$, $s = 4$, $g = 10$, $\alpha = 440$ *and* $G \cong \mathrm{aut}(M_{12})$,
(viii) $k = 4$, $s = 4$, $g = 10$, $\alpha = 880$ *and* $G \cong \mathrm{aut}(\hat{M}_{12})$,
(ix) $k = 5$, $s = 4$, $g = 9$, $\alpha = 17442$ *and* $G' \cong J_3$ *or*
(x) $k = 5$, $s = 4$, $g = 9$, $\alpha = 52326$ *and* $G'' \cong J_3$.

In each case, Γ is uniquely determined.

We consider more closely the case $s = (g-1)/2$ to illustrate the method of proof. We first show that $s = 4$ and that a certain subgroup of G acts as a transitive extension of $PGL_2(q)$ in its natural permutation representation.

By [**14**], it follows that $q \leq 4$. (When $s = (g-2)/2$, it is more difficult to bound q.) The conclusion of (3.1) allows us to write down a presentation for the subgroups G_x and $G_{\{x,y\}}$ in each of the remaining cases by suitably modifying the Steinberg relations (see [**3**]) for the corresponding group of Lie type (compare [**1**]). It then becomes possible to translate the hypothesis $s = (g-1)/2$ into an additional relation which in each case either causes collapse or defines a finite group from which we can reconstruct the graph Γ.

An exciting part of this work is the fact that in it the sporadic group J_3 (among other groups) arises as a solution to a natural geometrical problem. No properties of J_3 are needed in the proof (except to identify the group which the Todd-Coxeter algorithm produces as actually being J_3) and, in fact, this work could well have lead to the discovery of J_3 independent of other developments. It is also interesting to consider the role of computer technology in the proof. Coset enumeration is an essential step, and its application in the solution to this problem seems the most natural way to proceed. Nevertheless, it remains a mystery why coset enumeration produces just these few groups and not, say, infinitely many others. Perhaps there is an explanation for this fact related to the application of covering spaces in the theory of Buekenhout geometries.

There is a result which is, in a sense, the opposite of what we are talking about. Suppose Γ is finite and s-transitive with $s \geq 4$. Those circuits Δ whose pointwise stabilizer equals $G_{x_0,\ldots,x_s}$ for every s-path $(x_0,\ldots,x_s)$ lying on Δ play a special role in the theory of these graphs. (They are crucial to the construction of the generalized polygon $\tilde{\Gamma}$ in (3.1) and (3.2).) When $k > 3$, there is only one G-orbit of these circuits; we let t denote the length of any one of them. (Often, but not necessarily, we have $g = t$; in case (vii) of (4.1), for instance, $t = 12$ but $g = 10$.) Then $t = 2(s-1)$ if and only if Γ is a generalized $(s-1)$-gon, but [**5**]:

(4.2) Theorem: *For each $v \geq 2$ and for each pair $s \geq 4$ and q allowed by (3.1) above, there are infinitely many finite s-transitive graphs satisfying $(*)$ with $t = 2v(s-1)$.*

The proof of (4.2) is based on calculations in the fundamental group of the generalized $(s-1)$-gon $\tilde{\Gamma}$ associated with Γ in (3.1) above.

The hypothesis $s \geq (g-2)/2$ of (4.1) was motivated by the study of distance-transitive graphs. (Of the graphs (i)–(x) of (4.1), only the first three are actually distance-transitive.) There is an independent consideration which leads to this inequality, as we now explain. As indicated above (see (2.1)), the circuits of length $2n$ – called *apartments* – play a special role in the theory of generalized n-gons. Their setwise stabilizers are the groups N in the notion of a BN-pair of rank 2; we have $N/N \cap B \cong D_{2n}$, and any two edges of Γ lie on some $2n$-circuit. In the search for apartment-like structures in the geometries associated with the sporadic simple groups, the following modification of the notion of a BN-pair of rank 2 seems natural:

(4.3) Definition: *Let Γ be a connected, bipartite graph, let Δ be a circuit in Γ and let $\{x, y\}$ be an edge on Δ. Let G be a subgroup of* $\mathrm{aut}^\circ(\Gamma)$ *such that* $(G_x, G_y; G_{xy})$ *is one of the amalgams characterized in* (3.2) *above. Let $B = G_{xy}$, let N be the setwise stabilizer of Δ in G and suppose that neither $N \cap G_x$ nor $N \cap G_y$ lies in B. We say that Γ has the BNB-property if every two edges of Γ lie on Δ^a for some $a \in G$, or equivalently, if $G = BNB$.*

In [13], graphs with the BNB-property are classified:

(4.4) Theorem: *Suppose Γ satisfies the BNB-property with respect to some group G. Then Γ is either a generalized polygon or the 3-fold cover of the $Sp_4(2)$-generalized quadrangle as in case* (iii) *of* (4.1).

The first step in the proof of (4.4) is to show that the BNB-property implies that $s \geq (g-2)/2$, where s is maximal such that G acts locally s-transitively on Γ (see the table on p.98 of [4]). The proof then continues largely along the lines sketched out for the proof of (4.1).

Instead of substituting local s-transitivity for s-transitivity in the hypotheses of (4.1), we can try to relax the hypothesis $s \geq (g-2)/2$. Here the result (4.2) represents a fundamental obstruction. We are nevertheless motivated by the observation that the sporadic groups Ru and Th both act 4-transitively on graphs Γ of valency $k = 6$ (i.e. with $q = 5$) and $g = t = 12$. By (4.2), the hypotheses $g = t = 12$ does not suffice to characterize these graphs; additional properties of these graphs, which somehow involve circuits of odd length, must be considered. In both cases, too, the index of G_x in G is far too large to allow for coset enumeration. Only the Ru-graph has been successfully characterized [12]. The hypotheses in this characterization, though, are group theoretical conditions which do not have any natural geometrical interpretation. In particular, they include the existence of a subgroup $F \cong {}^2F_4(2)$ (much larger than $G_x \cong 5^2 : GL_2(5)$) whose cosets can be successfully enumerated. These hypotheses are all basic properties of Ru observed to hold in [10], so the coset enumeration yields a new existence proof for this group, but more should be possible.

There is another result [28] which has at least a superficial resemblance to (4.1):

(4.5) Theorem: *Let Γ be locally s-transitive, let $\{x, y\}$ be an edge of Γ and let $k = |\Gamma(x)|$ and $m = |\Gamma(y)|$. Then there is a function f of three variables such that if $s \geq (g-z)/2$ then $s \leq f(k, m, z)$.*

The proof of (4.5) uses some of the essential ideas behind (3.1) but does not depend on the classification of finite simple groups (or on any computer calculations). Combined with a result of A. A. Ivanov [9], it yields an elementary proof of the fact that for given $k \geq 3$, there are only finitely many distance-transitive graphs of valency k. This result is due to P. Cameron [2], who proved it, however, by using the Sims Conjecture, the only proof of which requires the classification of finite simple groups. It is interesting that the [9]–[28] proof of Cameron's theorem, which is about vertex-transitive graphs, requires consideration of graphs which are only locally s-transitive.

Suppose Γ is a connected graph and G is a subgroup of aut(Γ) acting transitively on the vertex set of Γ such that for each vertex x,

$$PSL_n(q) \leq G_x^{\Gamma(x)} \leq P\Gamma L_n(q)$$

(up to isomorphism) for some $n \geq 3$ and some prime power $q = p^m$. An important open problem in the local theory of s-transitive graphs is to determine the structure of the stabilizer G_x (or, almost as good, to find a bound for $|G_x|$ in terms of n and q) in this case. By [26], this is the only case of s-transitive graphs with $s \geq 2$ in which the structure of the stabilizers has not been determined. By [23], the graph Γ is s-transitive for $s = 2$ or 3. If $s = 3$ and $p \geq 5$, then the best possible bound has been determined in [27]. This result has recently been extended by V.I. Trofimov (personal communication) to the case $s = 2$, but the cases $p = 2$ and $p = 3$ remain unsolved. This problem is closely related to pushing-up problems for which solutions exist when $n = 3$.

The results (4.1) and (4.4) should be viewed as typical to a class of problems where the goal is to characterize geometries from certain types of local information. The ideas used in the proof of (4.1) and (4.4) have been recently modified to other situations to achieve similar results (see, for instance, [34] and [35]). These techniques combined with other aspects of the theory of amalgams seem to be particularly suited to examining the role of generalized polygons as the fundamental building blocks in the geometrical theory of the sporadic groups.

References:

1. N. L. Biggs, Presentations for cubic graphs, in *Computational Group Theory*, ed. M.D. Atkinson, Academic Press, 1984, pp.57-63.

2. P. J. Cameron, There are only finitely many distance-transitive graphs of given valency greater than two, *Combinatorica* **2** (1982), 9-13.

3. R. W. Carter, *Simple Groups of Lie Type*, John Wiley and Sons, New York, 1971.

4. A. Delgado and B. Stellmacher, Weak (B, N)-pairs of rank 2, in *Groups and Graphs: New Results and Methods*, Birkhäuser, Basel, Boston, 1985.

5. A. Delgado and R. Weiss, On certain coverings of generalized polygons, *Bull. London Math. Soc.* **90** (1989), 235-243.

6. P. Fong and G. M. Seitz, Groups with a (B, N)-pair of rank 2, I-II, *Invent. Math.* **21** (1973), 1-57, and **24** (1974), 191-239.

7. A. Gardiner, Arc transitivity in graphs, *Quart. J. Math. Oxford* **24** (1973), 399-407.

8. D. Goldschmidt, Automorphisms of trivalent graphs, *Ann. Math.* **111** (1980), 377-406.

9. A. A. Ivanov, Bounding the diameter of a distance-regular graph, *Soviet Math. Dokl.* **28** (1983), 149-152.

10. A. Rudvalis, A rank 3 simple group of order $2^{14} \cdot 3^3 \cdot 5^3 \cdot 7 \cdot 13 \cdot 29$, I, *J. Algebra* **86** (1984), 181-218.

11. J. P. Serre, *Trees*, Springer-Verlag, Berlin, Heidelberg, New York, 1980.

12. G. Stroth and R. Weiss, A new construction of the group Ru, *Quart. J. Math. Oxford* **40** (1989), to appear.

13. G. Stroth and R. Weiss, On graphs with the BNB-property, preprint.

14. M. Suzuki, Transitive extensions of a class of doubly transitive groups, *Nagoya Math. J.* **27** (1966), 159-169.

15. J. G. Thompson, Bounds for orders of maximal subgroups, *J. Algebra* **14** (1970), 135-138.

16. J. Tits, *Buildings of Spherical Type and Finite BN-Pairs*, Lecture Notes in Math. **386**, Springer-Verlag, Berlin, Heidelberg, New York, 1974.

17. J. Tits, Non-existence de certains polygones généralisés, I-II, *Invent. Math.* **36** (1976), 229-246, and **51** (1979), 267-269.

18. J. Tits, Moufang octagons and the Ree groups of type ${}^2F_4$, *Amer. J. Math.* **105** (1983), 539-594..

19. W. T. Tutte, A family of cubical graphs, *Proc. Cambridge Phil. Soc.* **43** (1947), 459-474.

20. W. T. Tutte, On the symmetry of cubic graphs, *Canad. J. Math.* **11** (1959), 621-624.

21. W. T. Tutte, *Connectivity in Graphs*, University of Toronto Press, Toronto, 1966.

22. R. Weiss, s-Transitive graphs, in *Algebraic Methods in Graph Theory*, Coll. Math. Soc. J. Bolyai **25**, Budapest, 1978, pp.827-847.

23. R. Weiss, Elations of graphs, *Acta Math. Acad. Sci. Hungar.* **34** (1979), 101-103.

24. R. Weiss, Groups with a (B, N)-pair and locally transitive graphs, *Nagoya Math. J.* **74** (1979), 1-21.

25. R. Weiss, The nonexistence of certain Moufang polygons, *Invent. Math.* **51** (1979), 261-266.

26. R. Weiss, The nonexistence of 8-transitive graphs, *Combinatorica* **1** (1981), 309-311.

27. R. Weiss, Graphs with subconstituents containing $L_n(q)$, *Proc. Amer. Math. Soc.* **85** (1982), 666-672.

28. R. Weiss, On distance-transitive graphs, *Bull. London Math. Soc.* **17** (1985), 253-256.

29. R. Weiss, Distance-transitive graphs and generalized polygons, *Arch. Math.* **45** (1985), 186-192.

30. R. Weiss, A characterization of the group $\hat{M}_{12}$, *Groups and Geometries* **2** (1985), 555-563.

31. R. Weiss, A characterization and another construction of Janko's group J_3, *Trans. Amer. Math. Soc.* **298** (1986), 621-633.

32. R. Weiss, Presentations for (G, s)-transitive graphs of small valency, *Math. Proc. Cambridge Phil. Soc.* **101** (1987), 7-20.

33. R. Weiss, On a theorem of Goldschmidt, *Ann. Math.* **126** (1987), 429-438.

34. R. Weiss, A characterization of the group Co_3 as a transitive extension of HS, *Arch. Math*, to appear.

35. R. Weiss and S. Yoshiara, A geometrical characterization of the groups Suz and HS, *J. Algebra*, to appear.

36. H. Wielandt, *Subnormal Subgroups and Permutation Groups*, Ohio State University Lecture Notes, Columbus, 1971.

Character Tables of Commutative Association Schemes

Eiichi Bannai*

The purpose of this paper is to discuss many examples of commutative association schemes and their character tables. Here we explain the intention of our present research direction as well as our recent results in [4,5, 7, 8, 9] in an informal way without referring to technical details. This paper was mostly written in June 1988. In submitting this final version in these Proceedings in June 1989, I did not try to update this paper systematically. Instead, in order to show some new development, I added several references at the end. This paper is aimed at mathematicians whose primary interest is in finite group theory or closely related areas. We should be very pleased if this paper provokes them to be interested in this subject and to join us in pursuing this research direction.

A preliminary version of this paper was published in Japanese in the Proceedings of Japan Math. Soc. Algebra Symposium held at Fukui in July 1987. Another survey paper [43] on the related subjects aiming at more general readers and putting more emphasis on the role of orthogonal polynomials, is to be published in the Proceedings of NATO-ASI on Orthogonal Polynomials and their Applications held at the Ohio State University, May 22-June 3, 1989. This paper and [43] will supplement each other to some extent.

1. Commutative Association Schemes, Multiplicity-Free Permutation Groups, and Their Character Tables

Let G be a finite group and let H be a subgroup of G. Then the group G acts naturally on the set of cosets $X = H\backslash G$. We denote all the orbits of G acting on the set $X \times X$ by $\mathcal{O}_0 = \{(x,x) | x \in X\}, \mathcal{O}_1, \ldots, \mathcal{O}_d$, which are in one-to-one correspondence with the set of double cosets $H\backslash G/H$ and also in one-to-one correspondence with the orbits of H on X. Let $R_i (i = 0, 1, \ldots, d)$ be the relations of X defined by

$$(x, y) \in R_i \Leftrightarrow (x, y) \in \mathcal{O}_i.$$

* Supported in part by NSF grant DMS-8703075.

Then the following conditions (1), (2), (3) and (4) are satisfied.

(1) $R_0 = \{(x,x) | x \in X\}$,
(2) $R_0 \cup R_1 \cup \cdots \cup R_d = X \times X$ and $R_i \cap R_j = \emptyset$ (if $i \neq j$),
(3) For each $i \in \{0, 1, \ldots, d\}$,

$$^tR_i (:= \{(y,x) | (x,y) \in R_i\}) = R_j \quad \text{for some} \quad j \in \{0,1,\ldots,d\}.$$

(4) For each fixed (ordered triple of) $i, j, k \in \{0, 1, \ldots, d\}$,

$$p_{ij}^k = |\{z \in X | (x,z) \in R_i \quad \text{and} \quad (z,y) \in R_j\}|$$

is constant whenever $(x,y) \in R_k$.

On the other hand, a combinatorial structure $\mathcal{X} = (X, \{R_i\}_{0 \leq i \leq d})$, a pair of a finite set X and a set of nonempty relations $R_i (i = 0, 1, \ldots, d)$ satisfying the above four conditions (1), (2), (3) and (4), is called an association scheme of class d. (This is equivalent to what is called a homogeneous coherent configuration in [19].) Let A_i be the adjacency matrix with respect to the relation R_i. Then $A_i A_j = \sum_{k=0}^{d} p_{ij}^k A_k$ by the condition (3), and so $A_0, A_1, \ldots, A_d$ generate a semi-simple algebra $\mathcal{A}$ (over $\mathbb{C}$) of dimension $d+1$. This algebra is called Bose-Mesner algebra (Hecke algebra or Centralizer algebra) of the association scheme $\mathcal{X}$. An association scheme is called commutative if $A_i A_j = A_j A_i$ for all i and j, i.e. if the algebra $\mathcal{A}$ is commutative.

When the association scheme $\mathcal{X}$ comes from the action of a group G on the set $X = H \backslash G$, the Hecke algebra $\mathcal{A}$ is commutative if and only if the permutation character π $(= 1_H^G)$ is multiplicity-free, i.e. π is decomposed into the direct sum of distinct irreducible characters of G. In this paper we are mostly concerned with the commutative case.

Now, let us suppose that an association scheme $\mathcal{X} = (X, \{R_i\}_{0 \leq i \leq d})$ is commutative. Then, by definition, the Hecke algebra (Bose-Mesner algebra) $\mathcal{A}$ is commutative, and $\mathcal{A}$ has a unique set of primitive idempotents $E_0 = \frac{1}{|X|} J, E_1, \ldots, E_d$ (where J is the matrix whose entries are all 1). Thus, the algebra $\mathcal{A}$ has two bases $A_0, A_1, \ldots, A_d$ and $E_0, E_1, \ldots, E_d$. We denote by P the base-change matrix, i.e.,

$$(A_0, A_1, \ldots, A_d) = (E_0, E_1, \ldots, E_d) \cdot P.$$

So the entries of the ith column of P are the eigenvalues of the matrix A_i. We call P the *character table* of the commutative association scheme $\mathcal{X}$. P is also called the first eigenmatrix of $\mathcal{X}$. The term "character table" will be justified from the following reason.

Let G be a finite group. Let $C_0 = \{1\}, C_1, C_2, \ldots, C_d$ be all the conjugacy classes of G. Define the relations $R_i (i = 0, 1, \ldots, d)$ by

$$(x,y) \in R_i \Leftrightarrow yx^{-1} \in C_i.$$

Then $\mathcal{X} = (G, \{R_i\}_{0 \le i \le d})$ becomes a commutative association scheme. This association scheme is called the group association scheme of G. This association scheme is also obtained from the action of the group $G \times G$ on the set $G : g^{(g_1,g_2)} = g_1^{-1} g g_2$, $(g_1, g_2) \in G \times G$. Let P be the character table of this commutative group association scheme of G, and let us define the matrix T by

$$T = \begin{pmatrix} f_0 & & & 0 \\ & f_1 & & \\ & & \ddots & \\ 0 & & & f_d \end{pmatrix} \cdot P \cdot \begin{pmatrix} \frac{1}{k_0} & & & 0 \\ & \frac{1}{k_1} & & \\ & & \ddots & \\ 0 & & & \frac{1}{k_d} \end{pmatrix}$$

where $k_i = |C_i|$, the sizes of the conjugacy classes, and f_i are the degrees of irreducible characters of G (since there is a natural bijection between primitive idempotents of $\mathcal{A}$ and the irreducible characters of G). Then T becomes *the character table* of the group G in the usual sense. The matrices P and T are different only by a normalization, and one determines the other completely and easily. It seems that, in our viewpoint, the matrix P is more natural and easy to handle than T. For example the entries of P are all algebraic integers for any commutative association scheme, but this may not be true for matrix T, though it is true for group association schemes.

The primitivity of an association scheme is defined as follows. An association scheme $\mathcal{X} = (X, \{R_i\}_{0 \le i \le d})$, not necessarily commutative, is primitive if and only if for all $i \ge 1$ the graph (X, R_i) is connected. In the example of association schemes coming from the action of group G on $X = H \backslash G$, the association scheme $\mathcal{X}$ is primitive if and only if the permutation group G on X is primitive (or equivalently H is a maximal subgroup of G). In the example of group association scheme of G, the association scheme $\mathcal{X} = (G, \{R_i\}_{0 \le i \le d})$ is primitive if and only if the group G is a simple group.

The real problem we want to address in the future, probably in a somewhat distant future, is the classification problem of primitive commutative association schemes (of large class numbers d). But we feel that more preparations will be necessary before that problem really takes off. So at the present stage, we are mainly interested in collecting many examples of (primitive) commutative association schemes (of large class numbers d), and in calculating the character tables of these known examples of association schemes.

We conclude this section by mentioning three remarks aimed at finite group theorists.

Remark 1. It may be difficult for pure finite group theorists to understand the need to consider association schemes. In fact, by using the classification of finite simple groups, it is quite realistic that all the multiplicity-free (primitive) permutation group G on $X = H\backslash G$ may be classified in the near future. However, it is our philosophy and conviction that we will be able to obtain far deeper understandings of the mathematical objects by considering them at the level of association schemes rather than considering them only at the level of groups. There are many interesting association schemes which do not come from multiplicity-free permutation groups (though usually they are related to each other). The theory will lose most fascinating beauty and reality if we confine ourselves to the case of permutation groups, though I agree that those are very important ones.

Remark 2. The character table of a (commutative) association scheme provides us with very useful information of the structure of the association scheme. For example all the structure constants p_{ij}^k are calculable from the character table. One might think that in the case the association scheme comes from the action of G on $X = H\backslash G$, it is sufficient to know (i) the character table of G, (ii) the character table of H and (iii) the decomposition of the permutation character $\pi = 1_H^G$. But the information of all these three is still not enough to determine the character table of the association scheme. Actually knowing the character table of the association scheme is equivalent to knowing all the (zonal) spherical functions of G on X. In addition to the information mentioned above, if we know how each double coset $HxH (x \in G)$ is splitted into conjugacy classes of G, then we can determine the character table of the association scheme.

Remark 3. The main reason why we are mostly concerned with the commutative case (at the present stage) is the technical difficulties in the general non-commutative case. If we include the non-commutative case at this stage, the technique involved will be too burdensome at this point. Also it seems that the commutative case forms a fairly closed universe, similar to the compact symmetric homogeneous spaces. Also for the study of finite (simple) groups through their group association schemes, the commutative case will be sufficient to consider.

2. Examples of Commutative Association Schemes

In this section we list four main sources of (primitive) commutative association schemes of large class numbers d. Note that we are not saying that these are the only sources. But these four sources seem to be very important sources, and it seems difficult to find examples of such association schemes which are not related to one of these sources. Actually it is difficult to find any examples of primitive (commutative) association scheme of large class numbers which are not related to permutation groups. (We remark that many examples of non-group-case commutative association schemes, i.e. R_i are not obtained as orbits of G on $X \times X$ with $X = H\backslash G$, are obtained as subschemes of known group-case commutative association schemes, see

Bannai [44] for more details. We can regard these non-group-case commutative association schemes not *directly* related to permutation groups, but still are *remotely* related to permutation groups.)

(1) Let G be a classical (or Chevalley) group. Then G naturally acts on an appropriate vector space over a finite field; this statement is obvious for classical groups but a bit delicate for exceptional groups (cf. [2,10] etc.). H is the stabilizer of a subspace of the vector space, and G acts on $X = H\backslash G$.

(2) Let G be a classical (or Chevalley) group, H an irreducible subgroup of G (on the vector space of which G acts naturally). G acts on $X = H\backslash G$. Examples for type A are as follows: (a) $G = GL(2n,q)$, $H = Sp(2n,q)$, (b) $G = GL(n,q^2)$, $H = GL(n,q)$, (c) $G = GL(n,q^2)$, $H = GU(n,q^2)$, (d) $G = GL(2n,q)$, $H = GL(n,q^2)$. There are many other examples for other types.

Remark. In the above examples these association schemes (or permutation groups) themselves are not necessarily primitive, but primitive association schemes are attached to them naturally and obviously in the same way as the simple group $PSL(n,q)$ is attached to $GL(n,q)$.

(3) The group association schemes of finite (simple) groups G, i.e., those obtained from G by using conjugacy classes or equivalently those obtained from the action of $G \times G$ on G, see the previous section.

(4) Let V be a vector space over $GF(q)$ and let G_0 be a subgroup of $GL(V)$ $(GL(V) = GL(n,q))$. $G = V \cdot G_0$ acts on V, i.e., $G = V \cdot G_0$, $H = G_0$ and G acts on $X = H\backslash G$.

As the above explanation is a bit too sketchy, let me add several comments to supplement it.

Remark 1. (a) In (1) and (2), sometimes we need to choose G to be the group extended by an (outer) automorphism group to get a multiplicity-free permutation group.

(b) In (1), if H stabilizes an *isotropic* subspace, then H is a parabolic subgroup. The action of G on $H\backslash G$ in that case is well studied and the character table (of the association scheme) is controlled by the character table of the corresponding action of the Weyl group W on the cosets $W_J\backslash W$, where W_J is the subgroup of W corresponding to the parabolic subgroup H (cf. the works of Tits, Curtis-Iwahori-Kilmoyer, etc. see e.g. [23, 12]). In (1) when does 1_H^G become multiplicity free is "almost" completely determined. (A. Cohen told me at the conference that this is completely known when H stabilizes an isotropic subspace, cf. [46]. There are still some undecided cases when H stabilizes a nonisotropic subspace, cf. [21].)

(c) In (1), there are several cases where G acts on $H\backslash G$ multiplicity-freely for the stabilizer H of a non-isotropic subspace. These examples and their character tables will be discussed later in this paper.

Remark 2. For (2), the list of all possible pairs of a classical group G and a maximal multiplicity-free subgroup is obtained by Inglis' thesis [21] (at least for G with dimensions at least 14). This list will be extended to other (almost) simple groups.

Remark 3. The association scheme in (3) is primitive if and only if G is a simple group. Therefore the primitive association schemes in (3) are determined by using the classification of finite simple groups. The character tables of association schemes in (3) are readily obtained from the group character tables, and vice versa. So the determination of the character tables in (3) is equivalent to the well studied problem of the determination of the character tables of finite simple groups, (cf. Lusztig [33], Atlas [11], etc.). In order to get an association scheme of this type, a weaker algebraic structure of loop or quasi group is sufficient. So, there may be some examples of primitive commutative association schemes which come from undiscovered simple loops or quasi groups. It is not known whether there are many simple non-group (i.e., non-associative) loops or quasi groups. (This statement is not exact. Jonathan Smith told me that many nonassociative simple loops are constructed by modifying simple groups, but they give the same association scheme which come from the groups. So the real question here is whether there are many simple non-group loops which give association schemes which are not group association schemes for any group G.) It is known (cf. [14], [31]) by using the classification of finite simple groups that Paige's simple Moufang loops of order $\frac{1}{(q-1,2)}q^3(q^4-1)$ are the only non-group simple Moufang loops. The character tables of these simple Moufang loops will be discussed later in this paper.

Remark 4. Since all the association schemes in (4) are commutative (for any G_0), there are lots of examples. It is not easy to classify all association schemes in (4), even those which are primitive. Even the classification problem of distance-transitive graphs is still unfinished in this case. Some typical examples are $G_0 = GO^{\pm}(2n, q), GO(2n+1, q)$, $GU(n, q^2)$, etc. The character tables of these cases have just been calculated by W. M. Kwok [30]. If we take G_0 to be a subgroup of the Singer cyclic group, then the calculations of the character tables are closely related to the theory of Gaussian sums and cyclotomic numbers (cf. [36]). There may exist some relatives of (4) by allowing V not to be an elementary abelian group, e.g., V to be a product of simple groups, or possibly homocyclic groups (cf. [32] for the latter case). From a group theoretical viewpoint, it will be very desirable to obtain a similar kind of theorem for multiplicty-free primitive permutation groups as in Praeger-Saxl-Yokoyama [37] which was for distance-transitive case.

Remark 5. The sources (1), (2), (3), (4) are not quite disjoint. For example, the pair $G = 0_{2n+1}(2^r) = Sp(2n, 2^r)$ and $H = 0^{\pm}_{2n}(2^r)$, is considered to be in (1) if we regard $G = 0_{2n+1}(2^r)$, and in (2) if we regard $G = Sp(2n, 2^r)$. The example of Paige's simple Moufang loops is regarded as an example of (1) where $0^{+}_{8}(q)$ acting on the non-isotropic points.

3. Calculation of Character Tables (Character Table of Paige's Simple Moufang Loops as an Introduction)

The purpose of the rest of this paper is to discuss the determination of the character tables of many of the commutative association schemes listed in the previous section as (1), (2), (3) and (4). Since the determination of the character tables of the association schemes in (3) is equivalent to the determination of the character tables of the corresponding finite (simple) groups, we will only discuss three cases: (1), (2) and (4).

A very interesting feature of the determination of these character tables is that, in most cases, there exists a smaller association scheme whose character table controls the character table of the (big) association scheme, in a similar way that the character table of a Weyl group controls the character table of the Chevalley group. This is the main theme of this paper, and we will show many results and conjectures on this theme in the rest of this paper.

In this section we will see this theme for the association scheme of Paige's simple Moufang loop. This example is nothing but a special case of more general results which will be discussed later (cf. Table 4 in this paper). But this is an important example for us because it was through this example that we reached the above theme of the existence of a smaller association scheme which controls the other.

Paige's simple Moufang loop $M(q)$. For any finite field $GF(q)$, let us set

$$\mathcal{M} = \{\begin{pmatrix} a & \alpha \\ \beta & b \end{pmatrix} | ab - \alpha \circ \beta = 1, a, b \in GF(q), \alpha, \beta \in GF(q)^3\}$$

and define the product of two elements by

$$\begin{pmatrix} a & \alpha \\ \beta & b \end{pmatrix}\begin{pmatrix} a' & \alpha' \\ \beta' & b' \end{pmatrix} = \begin{pmatrix} aa' + \alpha\circ\beta' & a\alpha' + b'\alpha - \beta \times \beta' \\ a'\beta + b\beta' + \alpha \times \alpha' & bb' + \beta\circ\alpha' \end{pmatrix}$$

where for two vectors $\alpha, \beta \in GF(q)^3, \alpha \circ \beta$ denotes the usual inner (dot) product, and $\alpha \times \beta$ denotes the usual outer (Kronecker) product. Then we have $|\mathcal{M}| = q^3(q^4-1)$. Let $Z = \{\begin{pmatrix} 1 & 0 \\ 0 & 1 \end{pmatrix}, \begin{pmatrix} -1 & 0 \\ 0 & -1 \end{pmatrix}\}$. Then the quotient $M(q) = \mathcal{M}/Z$ is a simple Moufang loop of $q^3(q^4-1)/(q-1,2)$ elements and called Paige's simple Moufang loop.

For any loop M, the group $Gr(M)$ is defined to be the transitive permutation group on M generated by the left and right translation of all the elements of M, i.e.

$$Gr(M) = \{L(x), R(x) \mid x \in M\}$$

where $L(x) : y \mapsto xy$ for $y \in M$, $R(x) : y \mapsto yx$ for $y \in M$. The orbits $\mathcal{O}_0, \mathcal{O}_1, \ldots \mathcal{O}_d$ of the action of $Gr(M)$ on $M \times M$ are identified with the conjugacy classes of the loop M; in other words, this can be a definition of the conjugacy classes of the loop M. If we regard $\mathcal{O}_i$ as the ith association relation on M, we get a *commutative* association scheme of class d which is called the association scheme of loop M, and so we can think of the character table of M, i.e. the P-matrix.

In what follows, we assume that $q = 2^r$, the odd q case is similarly treated, but slightly more complicated.

It is proved that Paige's simple Moufang loop $M(2^r)$ consists of $q+1$ conjugacy classes (i.e., $d = q$) and that these conjugacy classes are very analogous to the conjugacy classes of the group $PSL(2, 2^r)$; actually $PSL(2, 2^r)$ has also $q+1$ conjugacy classes. So, both the character tables, i.e., P-matrices of the association scheme of $M(2^r)$ and $PSL(2, 2^r)$ have the same size $(q+1) \times (q+1)$. Actually, there exist surprising similarities between these matrices.

The group character table T of the group $PSL(2,q), q = 2^r$

	C_0	C	elements of order divisors of $q+1$ C_ℓ $(1 \leq \ell \leq \frac{q}{2})$	elements of order divisors of$(q-1)$ C_m $(1 \leq m \leq \frac{q}{2} - 1)$
identity 1	1	1	$1 \ldots 1$	$1 \ldots 1$
Steinberg χ	q	0	$-1 \ldots -1$	$1 \ldots 1$
$(1 \leq r \leq \frac{q}{2})$ θ_r	$q-1$ $\vdots$ $q-1$	-1 $\vdots$ -1	$-(\sigma^{r\ell} + \sigma^{-r\ell})$	0
$(1 \leq s \leq \frac{q}{2} - 1)$ ψ_s	$q+1$ $\vdots$ $q+1$	1 $\vdots$ 1	0	$(\rho^{sm} + \rho^{-sm})$

where σ is a primitive $(q+1)$st root of 1, and ρ is a primitive $(q-1)$st root of 1.

TABLE 1

The character table P of the group association scheme of $PSL(2,q), q = 2^r$, is readily calculated from T, and is given as follows:

$$\begin{pmatrix} 1 & (q+1)(q-1) & q(q-1)\ldots q(q-1) & q(q+1)\ldots q(q+1) \\ 1 & 0 & -(q-1)\ldots-(q-1) & q+1\ldots q+1 \\ & & & \\ 1 & -(q+1) & & \\ \vdots & \vdots & -q(\sigma^{r\ell}+\sigma^{-r\ell}) & 0 \\ 1 & -(q+1) & & \\ & & & \\ 1 & q-1 & & \\ \vdots & \vdots & 0 & q(\rho^{sm}+\rho^{-sm}) \\ 1 & q-1 & & \end{pmatrix}$$

TABLE 2

The character table $\widetilde{P}$ of the association scheme of $M(q), q = 2^r$, is given as follows ([8]).

$$\begin{pmatrix} 1 & (q^3+1)(q^3-1) & q^3(q^3-1)\ldots q^3(q^3-1) & q^3(q^3+1)\ldots q^3(q^3+1) \\ 1 & q^2-1 & -q^2(q-1)\ldots-q^2(q-1) & q^2(q+1)\ldots q^2(q+1) \\ & & & \\ 1 & -(q^3+1) & & \\ \vdots & \vdots & -q^3(\sigma^{r\ell}+\sigma^{-r\ell}) & 0 \\ 1 & -(q^3+1) & & \\ & & & \\ 1 & q^3-1 & & \\ \vdots & \vdots & 0 & q^3(\rho^{sm}+\rho^{-sm}) \\ 1 & q^3-1 & & \end{pmatrix}$$

TABLE 3

The reader will notice that except for the second row which corresponds to the Steinberg character of $PSL(2, 2^r)$, each entry of $\widetilde{P}$ is obtained from the corresponding entry of P by *just replacing* q by q^3. The entries of the second row need to be modified so that the orthogonality relations of the matrix $\widetilde{P}$ are satisfied. So, it is possible to say that the character table $\widetilde{P}$ of the association scheme of the loop $M(q), q = 2^r$, is *controlled* by the character table P of the group association scheme of $PSL(2, 2^r)$, by the replacement $q \to q^3$.

As we have already noticed in Remark 5 in the previous section, the group $Gr(M(2^r))$ is identified with the multiplicity-free permutation group $G = 0_8^+(2^r)$ acting on the cosets $X = H\backslash G$ with $H = 0_7(2^r)$, where H is regarded as the stabilizer of a non-isotropic (projective) point. So the pair of G and H is an example of (1) in the list of the previous section and it will be natural to consider the general case of $G = 0_{2m}^+(q)$ acting on the cosets $X = H\backslash G$ with $H = 0_{2m-1}(q)$. As you may have expected, each entry of the character table $\widetilde{P}$ of this association scheme is, except for the second row, obtained from the corresponding entry of the character table P of the group association scheme of $PSL(2,q)$ by just replacing q by q^{m-1} (here we assume $q = 2^r$). The entries in the second row are easily determined by the orthogonality relations.

This phenomenon also occurs for other examples in (1) in a similar but slightly modified way. Exactly speaking, this occurs for those pairs of G and H in (1) for which H stabilizes a non-isotropic subspace. (As we have already mentioned, the case H stabilizes a non-isotropic subspace is already extensively studied, as H becomes a parabolic subgroup.) An interesting feature here is that sometimes not just the group association schemes of $PSL(2,q)$ but also some smaller association schemes related to $PSL(2,q)$ appear in an interesting way.

Here we collect the results ([4,5]) in the following table.

groups	spaces	controlling association schemes
(i) $GO^{\pm}_{2m}(q)$ ($O^{\pm}_{2m}(q)$, q odd)	nonisotropic points (half of nonisotropic points)	$PGL(2,q) \times PGL(2,q)/PGL(2,q)$ ($PSL(2,q) \times PSL(2,q)/PSL(2,q)$)
(ii) $O_{2m+1}(q)$	positive-type nonisotropic points	$PGL(2,q)/D_{2(q-1)}$
(iii) $O_{2m+1}(q)$	negative-type nonisotropic points	$PGL(2,q)/D_{2(q+1)}$
(iv) $U_m(q)$	nonisotropic points	$PGL(2,q)/Z_{q+1}$
(v) $Sp_{2m}(q)$	nonisotropic lines	$PGL(2,q)/Z_{q-1}$

TABLE 4

Remark 1. As it is mentioned in Inglis [21], there may possibly exist commutative association schemes attached to the action of orthogonal groups on the sets of non-isotropic (projective) lines. It is not yet decided whether there exist such commutative association schemes. I think that they exist and their character tables will be calculated in a similar fashion as in the above examples of Table 4.

Remark 2. The work of J. Soto-Andrade is closely related to these results. He calculated spherical functions of the case (i) in Table 4, in connection with dual reductive pairs and Weil representations. The result for (i) (and for odd q) can also be obtained through the work of Soto-Andrade.

Remark 3. The association scheme denoted by $PGL(2,q)/Z_{q-1}$ in Table 4 (v) will need some comments. Exactly speaking the action of $PGL(2,q)$ on the cosets $PGL(2,q)/Z_{q-1}$ is not multiplicity-free. But a bigger group $Z_2 \times PGL(2,q)$ acts transitively on the same cosets. $PGL(2,q)/Z_{q-1}$

means the commutative association scheme coming from the last multiplicity-free permutation group. This kind of convention of notation is also used for the notation $PSL(2,q) \times PSL(2,q)/PSL(2,q)$ in Table 4 (i) for $0^{\pm}_{2m}(q)/0_{2m-1}(q)$.

4. Calculation of Character Tables (Further Examples and Conjectures)

First we discuss the character table of the commutative association scheme obtained from the multiplicity-free permutation group $G = GL(2n,q)$ acting on the cosets $X = H\backslash G$ with $H = Sp(2n,q)$. We sometimes write this association scheme $GL(2n,q)/Sp(2n,q)$ for brevity.

Here let $G = GL(2n,q)$ and $H = Sp(2n,q)$ a subgroup of G. Klyachko [29] proved that 1_H^G is multiplicity-free (hence we have a commutative association scheme) and that the double cosets (hence the orbits of G on $X \times X$) are in one-to-one correspondence with the conjugacy classes of the group $GL(n,q)$. So the size of the character table $\widetilde{P}$ of the association scheme is equal to the size of the character table T of the group $GL(n,q)$ (hence the character table P of the group association scheme of $GL(n,q)$). It seems natural to suspect that there should be a close connection between the matrices $\widetilde{P}$ and P as we had seen for $0^{+}_{2m}(q)/0_{2m-1}(q)$ in the previous section. Actually this is the case. We explain this connection, following [7].

Theorem 1: (Bannai-Kawanaka-Song [7]). *The character table $\widetilde{P}$ of the association scheme $GL(2n,q)/Sp(2n,q)$ is obtained as follows.*

Step 1. Take the character table P of the group association scheme of $GL(n,q)$. Since the character table T of the group $GL(n,q)$ is determined by J. A. Green [17], the matrix P is readily obtained from T. All the entries of P are polynomials in q.

Step 2. In each entry of P, replace q by q^2.

Step 3. (First notation) The rows of P are parametrized by the irreducible characters of the group $GL(n,q)$. We divide irreducible characters of $GL(n,q)$ into blocks where each block consists of the irreducible characters with a same semi-simple part. The set of irreducible characters with a same semi-simple part is identified with the set of partitions $(m_1, m_2, \ldots), m_1 \geq m_2 \geq \ldots \geq 0, m_1 + m_2 + \ldots = m$, for a nonnegative integer $m \leq n$. A semisimple character is a character whose degree is relatively prime to q, and it corresponds to the partition $(m, 0, 0, \ldots)$ among the irreducible characters with that semi-simple part. Among the partitions of m, we introduce the partial order, called the strong order, by

$$(m_1, m_2, \ldots) \succ (m'_1, m'_2, \ldots) \Leftrightarrow \sum_{i=1}^{\ell} m_i \geq \sum_{i=1}^{\ell} m'_i \text{ for all } \ell.$$

(Algorithm for Step 3) Inside each block of rows, using the row orthogonality relations that must hold for $\widetilde{P}$, we apply Schmidt's orthonormalization according to the partial ordering defined above, starting with the row corresponding to the semi-simple character. The resulting matrix is the character table $\widetilde{P}$ of the association scheme $GL(2n,q)/Sp(2n,q)$. Note that the rows of $\widetilde{P}$ corresponding to semi-simple characters of $GL(n,q)$ are just obtained by replacing q by q^2 in the corresponding entries of the matrix P.

Theorem 1 justifies that we can say that the character table of the association scheme $GL(2n,q)/Sp(2n,q)$ is controlled by the character table of the group association scheme of $GL(n,q)$ by the replacement $q \to q^2$.

Remark 1. This example is regarded as a q-analogue of the multiplicity-free permutaton group G/H with $G = S_{2n}$ (the symmetric group of degree $2n$) and $H = W(B_n)$ (Weyl group of type B_n, i.e., the wreath product of Z_2 and S_n). According to N. Kawanaka, the process of decomposing $1_{W(B_n)}^{S_{2n}}$ (due to James [25]) exactly corresponds to Schmidt's orthonormalization in Step 3 of Theorem 1.

Remark 2. A similar process to the above three steps in Theorem 1 seems to work for a fairly wide classe of association schemes. Some of such examples will be discussed later in this section. In the case of the association scheme of Paige's simple Moufang loop $M(q)$, only the second row of $\widetilde{P}$ fails to be obtained by the mere replacement of q by q^3 from the corresponding row of the matrix P (see Section 3). This is because the second row, i.e., Steinberg character, is the only non-semisimple character of $PSL(2,2^r)$, and because the second row of $\widetilde{P}$ is obtained by Schmidt's orthonormalization of the rows of $\widetilde{P}$ after the replacement of q by q^3. That is, the first and the second rows consititute a block, and any other block consists of a single row.

Many examples (and possible examples) of a pair of a classical group G and its multiplicity-free maximal subgroup H are listed in Inglis [21]. For some of them we can conjecture what their character tables are. In some cases we have very precise conjectures and in some cases we have only vague conjectures at this stage. Let us discuss some of them here.

Gow [18] showed that for (a) $G = GL(n,q^2)$ and $H = GL(n,q)$ and (b) $G = GL(n,q^2)$ and $H = GU(n,q^2)$, the permutation character 1_H^G is multiplicity-free, and that the double cosets $H\backslash G/H$ are in one-to-one correspondence with the conjugacy classes of the group (a) $GU(n,q^2)$ and (b) $GL(n,q)$ respectively. This suggests that the character table $\widetilde{P}$ of the association scheme $GL(n,q^2)/GL(n,q)$ (respectively $GL(n,q^2)/GU(n,q^2)$)

is closely related to the character table P of the group association scheme of $GU(n, q^2)$ (respectively $GL(n, q)$). We propose the following:

Conjecture 2. The character table $\widetilde{P}$ of the association scheme $GL(n, q^2)/GL(n, q)$ (respectively $GL(n, q^2)/GU(n, q)$) is controlled by the character table P of the group association scheme of $GU(n, q^2)$ (respectively $GL(n, q)$) by the replacement $q \to -q$. Actually exactly the same process as in Theorem 1 is expected to hold with the Step 2 changed to the replacement of q by $-q$ (instead of by q^2).

We believe that Conjecture 2 can be generalized to other cases. The conjecture we make in the following is of a general nature, but a bit vague and premature. The exact formulations need to be worked out more precisely. (Please do not regard the statements in the next paragraph as a rigorous conjecture, but regard them as a first approximation to the reality.)

Let $G(q)$ be a (nontwisted Chevalley) group of any type (but of an appropriate kind, e.g., take $G(q) = GL(n, q)$ for type A_{n-1}). Let $G = G(q^2)$ and let $H = G(q)$. Then 1_H^G is expected to be "almost" multiplicity-free. (That is, though we may not necessarily get a multiplicity-free permutation group, we can get a commutative association scheme attached to this permutation group.) Suppose this is the case. Let us define

$$H^*(q) = \begin{cases} {}^2G(q), & \text{if the Weyl group } W \text{ of } G(q) \text{ has no element } -1, \\ & \text{i.e., the Dynkin diagram is of type } A_n, D_{2n+1} \text{ or} \\ & E_6, \\ G(q), & \text{otherwise (but possibly not of type } E_8). \end{cases}$$

(The above condition written after ${}^2G(q)$ was suggested by Kawanaka. My original condition was attributed to the existence of an involutive automorphism of the Dynkin diagram which does not change the length of the root. In view of Theorem 6.7 in Shoji [39], the above condition seems to be the right condition.)

Then the character table $\widetilde{P}$ of (the supposed) commutative association scheme G/H is "essentially" controlled by the characater table P^* of the group association scheme of $H^*(q)$ by the replacement $q \to -q$. Also the character table $\widetilde{P}^*$ of (the supposed) commutative association scheme G/H^* is "essentially" controlled by the character table P of the group association scheme of H by the replacement $q \to -q$. (For the example of $Sp(4, q^2)/Sp(4, q)$, see a discussion in the next section.)

We can also formulate a conjecture for the character table of the commutative association scheme $Sp(4, q)/Sz(q)$. A very exact conjecture on this character table is given in [9]. The outline of the conjecture is stated as follows. Let $q = 2^{2m+1}$. Write $2q = r^2$, i.e., $r = 2^{m+1}$. Then the entries of the character table P of the group association scheme of $Sz(q)$ are all expressed as polynomials in q and r. Then the character table $\widetilde{P}$ of the association scheme $Sp(4,q)/Sz(q)$ is controlled by the character table P of the group association scheme of $Sz(q)$ by the replacement $q \to q$ (fixed), $r \to -r$. (Exactly speaking, the step of applying Schmidt's orthonormalization is slightly needed to be modified, but this is a very similar process as in $0^{+}_{2m}(q)/0_{2m-1}(q)$ or $GL(2n,q)/Sp(2n,q)$. Character tables of $G_2(q)/Ree(q)$ and $F_4(q)/Ree(q)$ are also conjectured though not as precise as in the case of $Sp(4,q)/Sz(q)$ (assuming that we get commutative association schemes attached to these permutation groups), but we omit the details.

For the examples of association schemes in list (4) of Section 2, we get several interesting examples of smaller controlling association schemes. Typical ones are $G = V \cdot G0^{\pm}_{2m}(q), V \cdot G0_{2m+1}(q)$. They are investigated by Kwok [30] as well as other cases. Here we omit the description of the details of his results, but calculations very similar to those in [8, 4, 5] are involved. We also remark that in the calculations of the character tables of some association schemes in list (4), cyclotomic numbers and Gaussian sums are closely related (cf. [36]).

Remark 3. It is shown by Inglis [21] that for $G = GU(2n,q^2)$ and $H = Sp(2n,q)$, the permutation character 1_H^G is multiplicity-free. It is conjectured that the character table of this association scheme is controlled by the character table of the group association scheme of $GU(n,q^2)$ by the replacement $q \to q^2$.

Remark 4. For $G = GL(2n,q)$ and $H = GL(n,q^2)$, it is known by [22] that 1_H^G is multiplicity-free. We do not know what parametrizes the double-cosets $H\backslash G/H$.

5. Some Examples of the Decomposition of 1_H^G into Irreducible Characters

When the association scheme comes from the action of a group G on the cosets $X = H\backslash G$ for a subgroup H, to know the decomposition of the permutation character 1_H^G is not enough to determine the character table, as we have mentioned in Remark 2 in Section 1. However, this is an important first step, and is of independent interest from a group theoretical viewpoint. Here we will discuss some examples of the decompostions. First we will discuss three examples

(a) $GL(n,q^2)/GL(n,q)$,
(b) $GL(n,q^2)/GU(n,q^2)$,
(c) $GL(2n,q)/Sp(2n,q)$.

Let us recall the parametrization of the irreducible characters of the group $GL(n,q)$ due to Green [17], and the parametrization of the irreducible characters of the general unitary group $GU(n,q^2)$ by Ennola [15]. The parametrization by Ennola [15] was a conjecture (called Ennola conjecture) for a long time, but because of the recent development of the representation theory of finite Chevalley groups, it is now a theorem (cf. Hotta-Springer [20], Kawanaka [26]).

Let $\mathcal{F}$ be the set of all irreducible monic polynomials $f(x)$ with $f(x) \neq x$, i.e., $f(x) = x^d + a_1 x^{d-1} + \ldots + a_d$ with $a_d \neq 0$. For each $f \in \mathcal{F}$, let $\nu(f)$ be a partition of a nonnegative integer $|\nu(f)|$. Then the irreducible characters of the group $GL(n,q)$ are parametrized by the set of the following symbols (for the details, see [17]):

$$\prod_{f \in \mathcal{F}} f^{\nu(f)}$$

with $\sum_{f \in \mathcal{F}}$ (degree of $f(x)$) $\cdot |\nu(f)| = n$.

Now let $f(x) = x^d + a_1 x^{d-1} + \ldots + a_d$ with $a_d \neq 0$ be a monic polynomial with a_i in $GF(q^2)$. Then we write

$$\tilde{f}(x) = \overline{a}_d^{-1}(\overline{a}_d x^d + \overline{a}_{d-1} x^{d-1} + \ldots + 1)$$

where $\overline{a}_i$ denotes the conjugate of a_i by $Gal(GF(q^2)/GF(q))$. We say that a monic polynomial $g(x) \in GF(q^2)[x]$ is *U-irreducible*, if either $g(x)$ is irreducible and $g(x) = \widetilde{g(x)}$ or $g(x) = f(x)\widetilde{f(x)}$ where $f(x)$ is irreducible and $f(x) \neq \widetilde{f(x)}$. Note that $g(x)$ is U-irreducible if and only if $g(x) = \widetilde{g(x)}$ and $g(x)$ cannot be written in the form $g(x) = g_1(x)g_2(x)$ where $g_1(x)$ and $g_2(x)$ are nonconstant polynomials over $GF(q^2)$ such that $g_1(x) = \widetilde{g_1(x)}$ and $g_2(x) = \widetilde{g_2(x)}$.

Let $\mathcal{G}$ be the set of all irreducible monic polynomials $g \in GF(q^2)[x]$. For each $g \in \mathcal{G}$ let $\nu(g)$ be a partition of a nonnegative integer $|\nu(g)|$. Then the irreducible characters of the group $GU(n,q^2)$ are parametrized by the set of the following symbols (for the details, see [15]):

$$\prod_{g \in \mathcal{G}} g^{\nu(g)}$$

with $\sum_{g \in \mathcal{G}}$ (degree of $g(x)$) $\cdot |\nu(g)| = n$.

Now the decompositions of 1_H^G for the three cases (a), (b) and (c) are given as follows.

(a) $GL(n,q^2)/GL(n,q)$.

The irreducible characters of $GL(n,q^2)$ which appear in $1_{GL(n,q)}^{GL(n,q^2)}$ are exactly those obtained as follows:

For each irreducible character $\prod_{g \in \mathcal{G}} g^{\nu(g)}$ of the group $GU(n, q^2)$, assign the irreducible character of $GL(n, q^2)$ by replacing the expression as

$$g^{\nu(g)} = \begin{cases} g^{\nu(g)} & \text{if } g \text{ is an irreducible polynomial in } GF(q^2)[x] \\ g_1^{\nu(g)} g_2^{\nu(g)} & \text{if } g(x) = g_1(x)g_2(x) \text{ in } GF(q^2)[x]. \end{cases}$$

(b) $GL(n, q^2)/GU(n, q^2)$.
The irreducible characters of $GL(n, q^2)$ which appear in $1_{GU(n,q)}^{GL(n,q^2)}$ are exactly those obtained as follows:

For each irreducible character $\prod_{f \in \mathcal{F}} f^{\nu(f)}$ of the group $GL(n, q)$, assign the irreducible character of $GL(n, q^2)$ by replacing the expression as

$$f^{\nu(g)} = \begin{cases} f^{\nu(g)} & \text{if } f \text{ is an irreducible polynomial in } GF(q^2)[x] \\ f_1^{\nu(g)} f_2^{\nu(g)} & \text{if } f(x) = f_1(x)f_2(x) \text{ in } GF(q^2)[x]. \end{cases}$$

(c) $GL(2n, q)/Sp(2n, q)$. (cf. [7])

The irreducible characters of $GL(2n, q)$ which appear in $1_{Sp(2n,q)}^{GL(2n,q)}$ are exactly those obtained as follows:

For each irreducible character $\prod_{f \in \mathcal{F}} f^{\nu(g)}$ of the group $GL(n, q)$, assign the following irreducible character of $GL(2n, q)$

$$\prod_{f \in \mathcal{F}} f^{2\nu(g)},$$

where $2\nu(g)$ is a partition $(2m_1, 2m_2, \ldots)$ if $\nu(g)$ is a partition $(m_1, m_2, \ldots)$. (This result corresponds to the decomposition of $1_{W(B_n)}^{S_{2n}}$, where each of the irreducible characters of S_{2n} whose corresponding Young diagram (i.e. a partition of $2n$) *with each row length even* appears exactly once.

Remark 1. For $1_{GL(n,q^2)}^{GL(2n,q)}$, we do not know the exact decomposition at this time. It is proved in [38] that those appearing in $1_B^{GL(2n,q)}$, where B is a Borel subgroup of $GL(2n, q)$, are exactly those whose Young diagram have each row length even. This means that $|GL(n, q^2)\backslash GL(2n, q)/P_J| = |Sp(2n, q)\backslash GL(2n, q)/P_J|$ for any parabolic subgroup P_J of $GL(2n, q)$.

Remark 2. The complete decomposition of $1_{Sp(4,q)}^{Sp(4,q^2)}$ was determined by H. Yamada and Y. Iwakata and also by Lawther independently, by using the character tables of $Sp(4, q)$ obtained by Srinivasan (for odd q) and Enomoto (for even q). These permutation representations are not multiplicity-free, but the multiplicities are all *at most two.* In the case of $q = 2^r$, there is exactly one irreducible charcter, namely the reflection character of $Sp(4, 2^r)$, with multiplicity two. I hoped that if we consider $\widetilde{G} = Sp(4, q^2) \cdot \langle\sigma\rangle$ and $\widetilde{H} = Sp(4, q) \cdot \langle\sigma\rangle$ where σ is the field automorphism of $Sp(4, q^2)$ then $1_{\widetilde{H}}^{\widetilde{G}}$ is multiplicity-free. Unfortunately this is not the case. This implies that at the level of groups we cannot get multiplicity-free permutation groups attached to $Sp(4, q^2)/Sp(4, q)$ for odd q. Anyway, it is hoped that certain commutative association schemes are still related to this permutation group, and that it will be possible to calculate their character tables. To consider other small rank cases will be helpful for proving (or modifying) the conjectures on $G(q^2)/G(q)$ mentioned in Section 4.

6. Speculations and Future Research Problems

Our ultimate purpose in this research direction is to discuss the classification problem of primitive commutative association schemes of large class numbers. As the first step for this problem, we would like to study what are the possible character tables of such associaiton schemes. Of course this is not an easy problem at all, as the solution to this problem includes the possible character tables of multiplicity-free permutation groups and in particular the possible character tables of finite simple groups.

I believe that there is close analogy between the classification of primitive commutative association schemes and the classification of compact symmetric spaces, though the finite case is, as usual, far more difficult and complicated than the continuous case.

Knowing the character table of an association scheme corresponds to knowing the spherical functions of a compact symmetric space. Compact symmetric spaces are well studied. They are completely classified, and the spherical functions are all calculated. It is known that in a compact symmetric space there always exist a rank ℓ and a root system (of rank ℓ). The spherical functions are described by using the generalized Jacobi orthogonal polynomials of ℓ variables, whose orthogonality is defined by using the root system, (see e.g. [42] etc.). For example,

(a) $SU(n)$, i.e. $SU(n) \times SU(n)/SU(n)$,

(b) $SU(n)/S0(n)$,

(c) $SU(2n)/Sp(2n)$

are compact symmetric spaces attached to the root system of type A_{n-1}. The spherical functions are described by the generalized Jacobi orthogonal polynomials of $n-1$ variables, in which the weight functions defining the orthogonality are slightly different in each case.

Roughly speaking, the compact group $SU(n)$ corresponds to the finite group $GL(n, q)$ and the space $SU(2n)/Sp(2n)$ corresponds to the association scheme $GL(2n, q)/Sp(2n, q)$. So, it may not be very surprising that the representations of the association scheme $GL(2n,q)/Sp(2n,q)$ are analogous to those of the group $GL(n,q)$ because both are related to the root system of type A_{n-1}. Furthermore, the change $q \to q^2$ which appeared in Theorem 1 corresponds to the change of the weight functions which define the generalized Jacobi orthogonal polynomials of $SU(n)$ and $SU(2n)/Sp(2n)$.

The class of association schemes which are called P- and Q-polynomial association schemes has drawn serious attention recently, and the classification problem of such association schemes is successfully progressing (notably by P. Terwilliger, A. A. Ivanov, and others). This class of association schemes, roughly speaking, corresponds to the rank one compact symmetric spaces. An important feature of P- and Q-polynomial association schemes is that their character tables are described by Askey-Wilson discrete orthogonal polynomials (due to Leonard, cf. [6]). The natural question is what are the rank ℓ P- and Q-polynomial association schemes. Currently we do not have the right definition of them, but I believe that many of the (known) examples in the list of Section 2 should be in this class, and that there should be generalizations of Askey-Wilson discrete orthogonal polynomials to serveral variable. I believe that the character tables of rank ℓ P- and Q-polynomial association schemes should be described by using these generalized Askey-Wilson polynomials of ℓ variables. Of course there is room for debate whether general primitive commutative association schemes of large class numbers have nice properties in general, e.g. nice character tables readily describable by nice orthogonal polynomials of several variables, but it seems that at least those known examples have some nice properties like that. At this stage, it will be important and interesting to collect as many examples of commutative association schemes as possible and to calculate their character tables. In doing this, the principal theme of the existence of smaller controlling association schemes is, I believe, helpful to pursue this research direction further.

Remark 1. As noted in [6, p. 382], the simultaneity of the discovery of Askey-Wilson polynomials and the introduction of the concept of P- and Q-polynomial association scheme was quite interesting. For the higher rank case, something similar may happen again. I learned from several experts on orthogonal polynomials and special functions that (several kinds of) multivariable orthogonal polynomials which generalize the usual one variable Askey-Wilson polynomials are now discovered, and that they are at a center of interest. Askey-Wilson polynomials are the most general classes of orthogonal polynomials which contain all classical polynomials as either special or limiting cases, and are characterized by the property which is abstracted from the property of P- and Q-polynomial association schemes (due to Leonard, cf. [6]). I believe that the right class of generalized Askey-Wilson polynomials of several variables may be grasped and characterized again through the properties abstracted from some combinatorial properties.

Remark 2. The concept of generic rings was very useful to understand the relations between 1_B^G and 1_1^W, or $1_{P_J}^G$ and $1_{W_J}^W$ for a parabolic subgroup P_J of a Chevalley group G, (cf. [23], [12] etc.). It would be very interesting to know whether similar concepts are obtained in the course of understanding the connection between the character tables of commutative association schemes and the smaller controlling association schemes.

Acknowledgment. The development of the research direction outlined in this paper owes a great deal to many mathematicians. The author would like to thank them all, and in particular Noriaki Kawanaka who is really changing the phase of this research from infancy to real theory.

References:

1. M. Aschbacher: On the maximal subgroups of the finite classical groups, Invent. Math. 76(1984), 469–514.
2. M. Aschbacher: The 27-dimensional module for E_6, I, Invent. Math. 89(1987), 159–195.
3. B. Bagchi and N. S. N. Sastry: Intersection pattern of the classical ovoids in symplectic 3-space of even order, (preprint).
4. E. Bannai, S. Hao and S. Y. Song: Character tables of the association schemes of finite orthogonal groups acting on the nonisotropic points, to appear in J. of Combinatorial Theory(A).
5. E. Bannai, S. Hao, S. Y. Song, and H. Z. Wei: Character tables of the association schemes coming from finite unitary and symplectic groups, (preprint).
6. E. Bannai and T. Ito: Algebraic Combinatorics I, Benjamin/Cummings, Menlo Park, California, 1984.
7. E. Bannai, N. Kawanaka and S. Y. Song: The character table of the Hecke algebra $H(GL_{2n}(F_q), Sp_{2n}(F_q))$, to appear in J. of Algebra.
8. E. Bannai and S. Y. Song: The character tables of Paige's simple Moufang loops and their relationship to the character tables of $PSL(2, q)$. Proc. London Math.Soc. 58(1989), 209–236.
9. E. Bannai and S. Y. Song: On the character table of the association scheme $Sp(4, q)/Sz(q)$, to appear in Research Problem section of Graphs and Combinatorics.
10. A. M. Cohen and B. N. Cooperstein: The 2-spaces of the standard $E_6(q)$-module, Geom. Dedicata 25(1988), 467–480.

11. J. H. Conway, R. T. Curtis, S. P. Norton, R. A. Parker and R. A. Wilson: ATLAS of Finite Groups, Clarendon Press, Oxford, 1985.

12. C. W. Curtis, N. Iwahori and R. Kilmoyer: Hecke algebras and characters of parabolic type of finite groups with BN-pairs, I.H.E.S. Publ. Math. 40(1971), 81–116.

13. P. Delsarte: An algebraic approach to the association schemes of coding theory, Philips Research Rept Suppls, No. 10, 1973.

14. S. Doro: Simple Moufang loops, Math. Proc. Camb. Phil. Soc. 83 (1978), 377–392.

15. V. Ennola: On the characters of the finite unitary groups, Ann. Acad. Sci. Fenn. Ser Al No. 323(1963), 35 pp.

16. I. M. Gelfand and A. V. Zelevinskii: Models of representations of classical groups and their hidden symmetries, Functional Analysis and Its Applications 18(1984), 183–198.

17. J. A. Green: The characters of the finite general linear groups, Trans. Amer. Math. Soc. 80(1955), 402–447.

18. R. Gow: Two multiplicity-free permutation representations of the general linear group $GL(n, q^2)$, Math. Z. 188(1984), 45-54.

19. D. G. Higman: Coherent configurations, Geom. Dedicata 4(1975), 1–32.

20. R. Hotta and T. A. Springer: A specialization theorem for certain Weyl group representations and an application to the Green functions of the unitary groups, Invent. Math. 41(1977), 113–127.

21. N. F. J. Inglis: On multiplicity-free permutation representations of finite classical groups, Ph.D. thesis, Cambridge, 1988.

22. N. F. J. Inglis, M. W. Liebeck and J. Saxl: Multiplicity-free permutation representations of finite linear groups, Math. Z. (1986), 329–337.

23. N. Iwahori: Generalized Tits system (Bruhat decomposition) on p-adic semisimple groups, Proc. Symp. in Pure Math. vol. 9, (1966), 71–83.

24. A. T. James: Zonal polynomials of the real positive definite symmetric matrices, Ann. of Math. 74(1961), 456–469.

25. A. T. James: Distributions of matrix variates and latent roots derived from normal sample, Ann. Math. Statist. 35(1964), 475–501.

26. N. Kawanaka: Generalized Gelfand-Graev representations and Ennola duality, Algebraic Groups and related topics, Adv. Stud. Pure Math. vol. 6, Kinokuniya, Tokyo and North-Holland, Amsterdam, 1985, 175-206.

27. N. Kawanaka: Generalized Gelfand-Graev representations of exceptional simple algebraic groups over a finite field, I, Invent. Math. 84(1986), 575–616.

28. P. Kleidman and M. W. Liebeck: A survey of the maximal subgroups of the finite simple groups, Geom. Dedicata 25(1988), 375–389.

29. A. A. Klyachko: Models for the complex representations of the groups $GL(n,q)$, Math. USSR Sbornik 48(1984), 365–379.

30. W. M. Kwok: Character tables of association schemes of affine type, Ph.D. thesis, Ohio State University, 1989.

31. M. W. Liebeck: The classification of finite simple Moufang loops, Math. Proc. Camb. Phil. Soc. 102(1987), 33–47.

32. R. A. Liebler and R. A. Mena: Certain distance-regular digraphs and related rings for characteristic 4, J. of Combinatorial Theory (A), 47(1988), 111–123.

33. G. Lusztig: (Many papers on the representation theory of finite Chevalley groups), e.g., Characters of reductive groups over a finite field, Ann. of Math. Stud. vol. 107, Princeton Univ. Press, princeton NJ, 1984.

34. I. G. Macdonald: Symmetric functions and Hall polynomials, Oxford Univ. Press, Oxford, 1979.

35. I. G. Macdonald: Commuting differential operators and zonal spherical functions, in "Algebraic Groups, Utrecht 1986," Lecture Notes in Mathematics, No. 1271, Springer-Verlag, 1987.

36. R. J. McEliece and H. Ramsey, Jr.: Euler products, cyclotomy and coding, J. Number Theory 4(1972), 302–311.

37. C. E. Praeger, J. Saxl and K. Yokoyama: Distance-transitive graphs and finite simple groups, Proc. London Math. Soc. 55(1987), 1–21.

38. J. Saxl: On multiplicity-free permutation representations, in "Finite Geometries and Designs," London Math. Soc. Lecture Note Series, No. 49, Cambridge Univ. Press, Cambridge, 1981.

39. T. Shoji: Green functions of reductive groups over a finite field, in "The Arcata Conference on Representations of Finite Groups," Proc. Symp. in Pure Math. 47(1987), 289–301.

40. J. Soto-Andrade: Harmoniques spheriques sur un corps fini, C. R. Acad. Sci. Paris, t.272 (1970), 1642–1645.

41. J. Soto-Andrade: Harmoniques spheriques sur un corps fini, Notas de le sociedad de Mathematica de Chile, 3(1984), 4–82.

42. L. Vretare: Formulas for elementary spherical functions and generalized Jacobi polynomials, SIAM J. Math. Anal. 15(1984), 805–833.

43. E. Bannai: Orthogonal polynomials in coding thery and algebraic combinatorics, to appear in Proceedings of NATO-ASI on Orthogonal Polynomials and Their Applications.

44. E. Bannai: Subschemes of some association schemes, (preprint).

45. E. Bannai and S. Y. Song: The character table of commutative association scheme coming from the action of $GL(n,q)$ on non-incident point-hyperplane pairs, to appear in Hokkaido Math. J.
46. A. E. Brouwer, A. Cohen and A. Neumaier: Distance-regular graphs. Springer, 1989(to appear).
47. J. Hemmeter: A new family of distance-regular graphs, to appear.
48. A. A. Ivanov, M. E. Muzichuk and V. A. Ustimenko: On a new family of (P and Q)-polynomial schemes, to appear in Europ. J. Combinatorics.
49. G. Lusztig; Symmetric spaces over a finite field, to appear.
50. A. Munemasa: The splitting fields of association schemes, to appear.
51. P. Terwilliger: Leonard pairs and dual polynomial sequences, to appear.
52. P. Terwilliger: The incidence algebra of a uniform poset, to appear.
53. Z. Wan: Finite geometries and block designs, to appear in the Proceedings of R. C. Bose Memorial Conference, Calcutta, December, 1988.

Codes and Curves

J.W.P. Hirschfeld

1. Introduction

In 1981 Goppa [5] gave a construction of linear codes beginning from an algebraic curve over the finite field $GF(q)$. The parameters of these codes are restricted by the Riemann-Roch theorem. They also give asymptotic codes better than those previously known. Here we describe the construction and give some of the subsequent results for curves of genus zero and one. For other surveys, see [2], [7], [10], [16], [29], [30], [32], [35].

2. Linear codes

A *linear code* S is a subspace of $V_{n,q}$, the n-dimensional vector space over $GF(q)$, the finite field of q elements. A *codeword* is an element x of $V_{n,q}$ and is written

$$(x_1, x_2, \ldots x_n) \quad \text{or} \quad x_1 x_2 \ldots x_n;$$

its *length* is n. If $y = y_1 y_2 \ldots y_n$ then the *distance* of x from y is

$$\rho(x, y) = |\{i \mid x_i \neq y_i\}|.$$

The weight of x is

$$w(x) = \rho(x, 0).$$

The *minimum distance* of S is

$$d = d(S) = min_{x \in S \setminus \{0\}} w(x).$$

If k is the dimension of S, then the *information rate* of S is

$$R = k/n.$$

Its *relative distance* is

$$\delta = d/n.$$

S is called an $[n, k, d]$ – *code*.

The *Main Coding Theory Problem* is to find codes that maximize both R and δ: these aims are contradictory!

For any matrix A, let A^* be its transpose. A *generator matrix* G for S has as its rows the elements of a basis for S and so is a $k \times n$ matrix. A *parity check matrix* H of S is an $r \times n$ matrix ($r = n - k$) of rank r such that

$$Hx^* = 0, \forall x \in S.$$

The *dual code* of S is $S^\perp = \{y \mid yx^* = 0, \quad \forall x \in S\}$. A code S' is *equivalent* to S if there exists some fixed element σ of the symmetric group $\mathbf{S}_n$ such that

$$S' = \{x\sigma = x_{1\sigma}x_{2\sigma}\ldots x_{n\sigma} \mid x \in S\}.$$

Theorem 2.1: *$d(S) = d$ if and only if every $d-1$ columns of H are linearly independent but some d columns are linearly dependent.*

Proof: There exists x in S of weight w if and only if $Hx^* = 0$, which in turn occurs if and only if some w columns of H are linearly dependent. ∎

Corollary: $n - k \geq d - 1$.

Proof: The rank of H is at least $d - 1$. ∎

When equality holds the code is called *optimal* or MDS (maximum distance separable).

Example: *Extended Reed-Solomon code.*
Let $\mathcal{P}_k = \{f \in GF(q)[X] \mid \deg f < k\}$ and let $GF(q) = \{t_1, \ldots, t_q\}$. Then

$$S = \{(f(t_1), \ldots, f(t_1)) \mid f \in \mathcal{P}_k\}.$$

Here $n = q$. Since any f has at most $k-1$ zeros, $d = q-(k-1) = n-k+1$. Since a basis for $\mathcal{P}_k$ is $1, X, X^2, \ldots X^{k-1}$, a suitable G is

$$\begin{bmatrix} 1 & 1 & \ldots & 1 \\ t_1 & t_2 & \ldots & t_q \\ \vdots & \vdots & & \vdots \\ t_1^{k-1} & t_2^{k-1} & \ldots & t_q^{k-1} \end{bmatrix}.$$

When q is even, the dual code can be further extended for $n = 2$. Finite geometers will recognize a regular (hyper) oval.

$$H = \begin{bmatrix} 1 & 1 & \ldots & 1 & 0 & 0 \\ t_1 & t_2 & \ldots & t_q & 1 & 0 \\ t_1^2 & t_2^2 & \ldots & t_q^2 & 0 & 1 \end{bmatrix};$$

$n = q + 2, r = d - 1 = 3$. Hence this is a $[q + 2, q - 1, 4]$-code.

There are several bounds on the parameters of a code.

Theorem 2.2: (*Sphere packing bound*)
If S is an $[n,k,d]$ - code over $GF(q)$, then

$$1+(q-1)\binom{n}{1}+\ldots+(q-1)^e\binom{n}{e}\leq q^{n-k},$$

where $e=[(d-1)/2]$.

Proof: The code S can correct e errors. For each x in S, there are $\binom{n}{i}$ ways of choosing i coordinates from x and hence $(q-1)^i\binom{n}{i}$ codewords with precisely i coordinates different from x. Hence there are $M=\sum_{i=0}^{e}(q-1)^i\binom{n}{i}$ vectors at distance up to e from x. Therefore $q^kM\leq q^n$. ∎

Theorem 2.3: (*Gilbert-Varshamov bound*)
There exists an $[n,k,d]$ - code over $GF(q)$ providing

$$1+(q-1)\binom{n-1}{1}+\ldots+(q-1)^{d-2}\binom{n-1}{d-2}<q^{n-k}.$$

Proof: It suffices to construct an $r\times n$ matrix with no $d-1$ columns linearly dependent. The first column can be any non-zero r-tuple. Now proceed by induction. Suppose the first i columns have been chosen so that no $d-1$ are linearly dependent. The sum of the number of distinct linear combinations of the columns taken one at a time, then two at a time and so on up do $d-2$ at a time is

$$N=(q-1)\binom{i}{1}+(q-1)^2\binom{i}{2}+\ldots+(q-1)^{d-2}\binom{i}{d-2}.$$

Provided $N<q^r-1$, another column may be added so that no $d-1$ columns of the augmented $r\times(i+1)$ matrix are linear dependent. This is the required inductive step. Now $i=n-1$ gives the result.

Lemma 2.4: *For any positive integer $t\leq n$,*

$$\frac{1}{n+1}q^{nH_q(t/n)}\leq\sum_{i=0}^{t}\binom{n}{i}(q-1)^i\leq q^{nH_q(t/n)},$$

where

$$H_q(X)=X\log_q(q-1)-X\log_q X-(1-X)\log_q(1-X).$$

Proof: For any real z such that $0 < z \leq 1$,

$$\sum_{i=0}^{t} \binom{n}{i} (q-1)^i \leq \sum_{i=0}^{t} \binom{n}{i} (q-1)^i z^{i-t}$$
$$\leq \sum_{i=0}^{n} \binom{n}{i} (q-1)^i z^{i-t}$$
$$= z^{-t}[1 + (q-1)z]^n.$$

The minimum value of this function of z gives the upper bound required. A similar argument gives the lower bound. Let

$$V_q = \{(R(S), \delta(S)) \mid S \text{ is a } q-\text{ary code}\}$$
$$U_q = \text{set of limit points of } V_q.$$

From these results, one deduces the following, also known as the Gilbert-Varshamov bound.

Theorem 2.5: *There exist a continuous function $\alpha_q(\delta)$ decreasing in the interval $[0, (q-1)/q]$ such that $U_q = \{(R, \delta) | 0 \leq R \leq \alpha_q(\delta)\}$, where*

$$\alpha_q(\delta) \geq 1 - H_q(\delta).$$

This result says that as n goes to infinity, there is a sequence of codes for which the upper limit of R is bounded below by $1 - H_q(\delta)$.

3. Algebraic curves and rational functions

Let F be a ternary form over $K = GF(q)$ and let $V(F)$ be the set of zeros of F in $PG(2, q)$, the projective plane over K. Let $\bar{K}$ be the algebraic closure of K and let (F) be the ideal in $K[X, Y, Z]$ generated by F.

Let $C = (V(F), (F))$; this is a *plane, projective, algebraic curve over* $GF(q)$. The elements of $V(F)$ are the *points of* C *rational over* $GF(q)$ or simply, *the points of* C, if there is no ambiguity. The *points of* C *rational over* $GF(q^r)$ are the zeros of F in $PG(2, q^r)$. It should be noted that $V(F)$ does not uniquely define F. For example, when $q = 4$ and $GF(4) = \{0, 1, \omega, \omega^2\}$, let $F = X^3 + \omega Y^3 + \omega^2 Z^3$. Then $V(F) = \{(1, y, z) | y^3 = z^3 = 1\} = V(F')$ where $F' = X^3 + \omega^2 Y^3 + \omega Z^3$. See [12] for the classification of cubics with 9 points for which there is a distinct cubic with precisely the same set of zeros.

Denote the set of points of C rational over $\bar{K}$ by $\bar{V}(F)$.

A *singular point* of C is a point (x, y, z) of C rational over $\bar{K}$ such that

$$\frac{\partial F}{\partial X} = \frac{\partial F}{\partial Y} = \frac{\partial F}{\partial Z} = 0 \text{ at } (x, y, z).$$

For example, let

$$F = (X^2 + Y^2)^2 + (X^2 - Y^2)Z^2 + Z^4.$$

If $q \equiv 1 \pmod 4$, then C has two singular points $(i, 1, 0)$ and $(i, -1, 0)$ rational over $GF(q)$, where $i^2 = -1$. If $q \equiv -1 \pmod 4$, then C has the same two singular points rational over $GF(q^2)$.

Let $P = (0, 0, 1)$ and suppose

$$F(X, Y, 1) = F_s + F_{s+1} + \ldots + F_m$$

where $F_s \neq 0$ and F_j is a form of degree j. Then P is a point of *multiplicity* s, providing $s \geq 1$; it is singular if $s > 1$ and an *ordinary* multiple point if F_s has no repeated factors. The factors of F_s over K are the *tangents* at P. Each curve C can be transformed by an invertible polynomial map to one with only ordinary multiple points. If C is such a curve with ordinary multiple points $P_1, \ldots, P_t$ of respective multiplicities $s_1 \ldots, s_t$, then the *genus* of C is

$$g = g(C) = \frac{1}{2}(m-1)(m-2) - \sum_{i=1}^{t} s_i(s_i - 1)/2.$$

If C has genus 0, it is *rational*; if it has genus 1, it is *elliptic*.

Let C be non-singular and let $\{P_1, \ldots, P_n\} \subset \bar{V}(F)$. Then $D = \Sigma n_i P_i$, $n_i \in \mathbf{Z}$, is a *divisor* of *degree* Σn_i. The divisor D is *effective*, written $D > 0$, if $n_i \geq 0$ all i. A *rational function* on C is $f = A/B$ where $A, B \in K[X, Y, Z]$. Here $f = f'$ if $f - f'$ is divisible by F. Crudely, the intersections of $V(A)$ with $V(F)$ give the *zeros* of f and the intersections of $V(B)$ with $V(F)$ give the *poles* of f. Associated to the zeros is the divisor D_A and to the poles D_B. This gives the divisor

$$\text{div } f = D_A - D_B$$

with $\deg D_A = \deg D_B$.

To determine the order r of P in D_A, write $f = u\ell^r$, where ℓ is linear and not a tangent at P, with $u(P) \neq 0, \ell(P) = 0$. For P in D_B, write $1/f = u\ell^r$; then r is the order of P in D_B. For example, suppose

$$F = Y^2 Z - X(X - Z)(X - \lambda Z), \ \lambda^2 \neq \lambda.$$

If $f = Y/Z$, then its zeros are $(0, 0, 1), (1, 0, 1), (\lambda, 0, 1)$ and its pole is $(0, 1, 0)$. Now,

$$\begin{aligned} \frac{1}{f} = \frac{Z}{Y} &= \frac{X(X - Z)(X - \lambda Z)}{Y^3} \\ &= X^3 \frac{1}{Y^3}(1 - \frac{Z}{X})(1 - \lambda \frac{Z}{X}) \\ &= X^3 \frac{1}{Y^3}\{1 - \frac{(X - Z)(X - \lambda Z)}{Y^2}\}\{1 - \frac{\lambda(X - Z)(X - \lambda Z)}{Y^2}\}. \end{aligned}$$

Hence, $(0,1,0)$ is a pole of order 3, and

$$\text{div } f = (0,0,1) + (1,0,1) + (\lambda,0,1) - 3(0,1,0).$$

For an effective divisor $D = \Sigma n_i P_i$, let

$$\begin{aligned} L(D) &= \{f | \text{div} f \geq -D\} \\ &= \text{set of rational functions } f \text{ such that div } f \\ &\quad \text{has poles of order no more than } n_i \text{ at } P_i. \end{aligned}$$

Theorem 3.1: (Riemann-Roch)
(a) *$L(D)$ is a vector space of dimension $\ell(D)$ over K.*
(b) *There exists g in $\mathbf{Z}$ such that*

$$\ell \leq \text{deg } D + 1 - g.$$

The smallest such g is the genus and $g \geq 0$.
(c) *If $\text{deg } D > 2g - 2$, then*

$$\ell(D) = \text{deg } D + 1 - g.$$

For example, consider the curve in Section 2 given by $F = X^3 + \omega Y^3 + \omega^2 Z^3$ with base field $K = GF(4)$. Its nine points $P_i, i = 1, \ldots, 9$, have as coordinates the columns of Table 1.

Table 1

P_1	P_2	P_3	P_4	P_5	P_6	P_7	P_8	P_9
1	1	1	1	1	1	1	1	1
1	1	1	ω	ω	ω	ω^2	ω^2	ω^2
1	ω	ω^2	1	ω	ω^2	1	ω	ω^2

Let $D = Q_1 + Q_2 + Q_3$, where $\{Q_1, Q_2, Q_3\} = \bar{V}(F) \cap \bar{V}(X)$. Then $\ell(D) = 3 + 1 - 1 = 3$. A basis for $L(D)$ is $f_1 = 1, f_2 = (X+Y)/X, f_3 = (X+Z)/X$.

4. Construction of geometric codes

Let $C = (V(F), (F))$ be a non-singular, projective, plane curve over $GF(q)$. Let $P_1, P_2, \ldots, P_n$ be distinct points of $V(F)$ with divisor $D = \Sigma P_i$. Let $E = \Sigma m_i Q_i$ be the intersection divisor of C with some other curve over $GF(q)$ and let $\text{deg } E = \Sigma m_i = m$; thus each Q_i is rational over $GF(q^r)$ for some r. Suppose further that $Q_i \neq P_j$, all i, j.

Consider the map

$$\vartheta : L(E) \to GF(q)^n$$

given by

$$\vartheta(f) = (f(P_1), \ldots, f(P_n)).$$

Then $S = S(D, E)$ = image of ϑ is the *geometric code defined on C by D and E*.

Theorem 4.1: *Let $S(D, E)$ have dimension k and minimum distance d. If $n > m > 2g - 2$, then* (a) $k = m - g + 1$; (b) $d \geq n - m$.

Proof: (a) Theorem 3.1 gives the result by putting $D = E$ and $k = \ell(E)$.

(b) If $f \in L(E)$ and $w(\vartheta(f)) = d$ then f is zero at $n - d$ points P_i forming the divisor D', where deg $D' = n - d$ and $D' < D$. So

$$\text{div } f > D' - E,$$

whence

$$\text{deg div} f \geq \text{deg } D' - \text{deg } E;$$

that is,

$$0 \geq n - d - m. \quad \blacksquare$$

Corollary 1: (a) $R + \delta \geq 1 - (g - 1)/n$; (b) $n - k + 1 \geq d \geq n - k + 1 - g$; (c) $n - m + g \geq d \geq n - m$.

Let the dual code $S(D, E)^\perp$ have dimension $k^\perp$ and minimum distance $d^\perp$.

Corollary 2: *If $n > m > 2g - 2$, then*
(a) $k^\perp = n - m + g - 1$;
(b) $m - g + 2 \geq d^\perp \geq m - 2g + 2$.

As an example, let us continue with the curve at the end of Section 3. As in Table 1, let $D = \sum_{i=1}^{9} P_i, E = Q_1 + Q_2 + Q_3 = \bar{V}(F) \cap \bar{V}(X)$.

For each point P_i the columns of the matrix G give the values of the functions f_1, f_2, f_3.

$$G = \begin{bmatrix} 1 & 1 & 1 & 1 & 1 & 1 & 1 & 1 & 1 \\ 0 & 0 & 0 & \omega^2 & \omega^2 & \omega^2 & \omega & \omega & \omega \\ 0 & \omega^2 & \omega & 0 & \omega^2 & \omega & 0 & \omega^2 & \omega \end{bmatrix}$$

Then G is the generator matrix of an $[n, k, d]$-code with $n = 9$, $k = 3$ and $7 \geq d \geq 6$. However, since two rows of G have weight 6, so $d = 6$. Thus $S(D, E)$ is a $[9, 3, 6]$-code. The dual code $S(D, E)^\perp$ is a $[9, 6, 3]$-code.

For $q = p^{2h}$, there exists a sequence of curves, called Shimura curves, for which $\beta = \lim_{n \to \infty} g/n = (\sqrt{q} - 1)^{-1}$.

Theorem 4.2: *For $q \geq 49$, with $\beta = (\sqrt{q}-1)^{-1}$, the line $R = 1 - \beta - \delta$ meets the curve $R = 1 - H_q(\delta)$ in two points δ_1 and δ_2. Thus there exists an infinite series of q-ary codes lying above the Gilbert-Varshamov bound.*

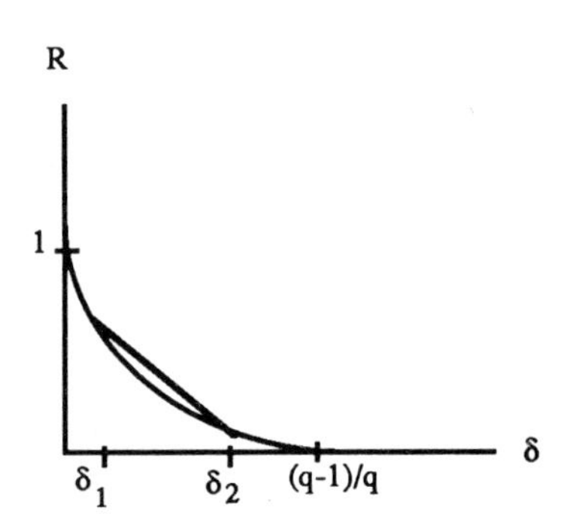

The Shimura curves $X_0(N)$ are derived from the action of

$$\Gamma_0(N) = \left\{ \begin{bmatrix} a & b \\ c & d \end{bmatrix} \in SL(2,\mathbf{Z}) \mid c \equiv 0 (\text{mod } N) \right\}$$

on the complex upper half-plane by $z \to (az+b)/(cz+d)$.

The curves $X_0(N)$ can be obtained as follows [32]. Let N be a prime other than p. The classical modular invariant J is given by

$$J(z) = \frac{(720)^3 \{\Sigma(mz+n)^{-4}\}^3}{(2\pi)^{12} e^{2\pi i z} \Pi(1 - e^{2\pi i s z})^{24}},$$

where the sum is taken over all (m,n) in $\mathbf{Z}^2 \backslash \{(0,0)\}$ and the product over all s in $\mathbf{N}$. Here J is defined on the upper half-plane. Let $J_N = J(Nz)$ and $J = J(z)$. There is a polynomial Φ_N in $\mathbf{Z}[X,Y]$ such that

$$\Phi_N(J, J_N) = 0.$$

Then $X_0(N)$ is the non-singular projective curve given by the polynomial Φ_N with its coefficients reduced mod p. If g is its genus and n the number of its points rational over $GF(p^2)$, then

$$g = [N/12],$$
$$|n - N(p-1)/12| \leq c_p$$

where c_p is independent of N. So as N and therefore n tends to infinity, $n/g \to p - 1 = \sqrt{q} - 1$.

Let $N_q(g)$ be the maximum number of points of any non-singular projective curve of genus g defined over $GF(q)$ and let $A_q = \limsup_{g\to\infty} N_q(g)/g$.

Theorem 4.3:
(a) (Vladut-Drinfeld) $A_q \le \sqrt{q} - 1$.
(b) *For q square, equality holds in* (a).

For the extremely elegant proof of (a) see [33] or [10]. Part (b) follows from the existence of the curves $X_0(N)$.

5. Rational curves

Every curve of genus 0 is projectively equivalent to a projective line or equally well a conic. Theorem 4.1 becomes the following.

Theorem 5.1: *For a geometric code $S(D,E)$ on the curve C of genus zero,*
(a) $0 \le m \le n-2 \Rightarrow k = m+1$ *and* $d = n - m$;
(b) $m \ge 0 \Rightarrow d = n - k + 1 \Rightarrow S(D,E)$ *is an MDS code.*

The generator matrix G of such codes can also be determined.

Theorem 5.2: (a) *If $n \le q$ there exists a generator matrix of the form*

$$G = \begin{bmatrix} 1 & \dots & 1 \\ \alpha_1 & \dots & \alpha_n \\ \vdots & & \vdots \\ \alpha_1^{k-1} & \dots & \alpha_n^{k-1} \end{bmatrix} \begin{bmatrix} t_1 & & & 0 \\ & t_2 & & \\ & & \ddots & \\ 0 & & & t_n \end{bmatrix}$$

Thus the code is a generalized Reed-Solomon code.

(b) *If $n = q+1$, there exists a generator matrix of the form*

$$G' = \begin{bmatrix} 1 & \dots & 1 & 0 \\ \alpha_1 & \dots & \alpha_q & 0 \\ \vdots & & \vdots & 0 \\ \alpha_1^{k-1} & \dots & \alpha_q^{k-1} & 1 \end{bmatrix} \begin{bmatrix} t_1 & & & & 0 \\ & t_2 & & & \\ & & \ddots & & \\ & & & t_q & \\ 0 & & & & 1 \end{bmatrix}$$

So the code is a projective generalized Reed-Solomon code.

For the equivalence of the codes $S(D,E)$ on C, we require the following definition: $E \underset{D}{\approx} E'$ if there is a rational function f such that $\operatorname{div} f = E - E'$ and $f(P_i) = 1$, all i; here $D = \Sigma P_i$.

Theorem 5.3: *The following are equivalent when* $1 \leq \deg E \leq n-3$:
(a) $S(D,E) = S(D',E')$ *with* $D = \Sigma P_i$, $D' = \Sigma P'_i$;
(b) *there exists* σ *in* $PGL(2,q)$ *such that* $\sigma(P) = P'$ and $\sigma(E) \underset{D'}{\approx} E'$.

To describe when $S(D,E)$ is cyclic, it is convenient to give a different description. Let

$$S_0 = S(D,E)^{\perp} = \{\mathbf{c} = (c_1, \ldots, c_n) \in GF(q)^n \mid \sum \frac{c_i}{z - \alpha_i} \equiv 0 \mod g(z)\}$$

and extend S_0 with a parity check to

$$\hat{S}_0 = \{(\mathbf{c}, c_{n+1}) \mid \mathbf{c} \in S_0, \sum_{i=1}^{n+1} c_i = 0\}.$$

Theorem 5.4: *In Theorem* 5.2(a), *let* $t_i = g(\alpha_i)^{-1}$ *for some polynomial* $g(z)$ *of degree* t. *Then*

$$S_0 = S(D,E)^{\perp},$$

where $E = E_0 - (t+1)P_\infty$, div $g = E_0 - tP_\infty$, *and* P_∞ *is the pole of* z. *A basis for* $L(E)$ *is*

$$1/g, z/g, \ldots z^{t-1}/g.$$

Theorem 5.5: (Stichtenoth [25]) *If one of the following holds, then* $\hat{S}_0$ *is cyclic:*
(a) $n+1 | q-1$ *and* $g(z) = (z-\alpha_1)^{r_1}(z-\alpha_2)^{r_2}, \alpha_1 \neq \alpha_2$;
(b) $n+1 | q+1$ *and* $g(z) = (z^2+bz+c)^r$, *where the quadratic is irreducible;*
(c) $n+1 = p$ *and* $g(z) = (z-\alpha)^r$.

6. Elliptic curves

Any elliptic curve over $GF(q)$ is isomorphic to a non-singular plane cubic curve C. Let C have N_i points rational over $GF(q^i)$. Then the zeta function on C is

$$\begin{aligned}\zeta(T) &= \exp(\sum N_i T^i / i) \\ &= \frac{1 - cT + qT^2}{(1-T)(1-qT)} \\ &= \frac{(1-\alpha T)(1-\bar{\alpha} T)}{(1-T)(1-qT)}\end{aligned}$$

where c is an integer and $|\alpha| = \sqrt{q}$. Hence

$$N_i = q^i + 1 - (\alpha^i + \bar{\alpha}^i),$$

which in turn implies that

$$\begin{aligned} N_1 &= q + 1 - c, \\ N_2 &= (q+1)^2 - c^2, \\ N_3 &= q^3 + 1 + 3qc - c^3, \\ N_4 &= (q^2-1)^2 + 4qc^2 - c^4. \end{aligned}$$

Primarily

$$q + 1 - 2\sqrt{q} \le N_1 \le q + 1 + 2\sqrt{q}.$$

The precise numbers that N_1 can be is given by the next result.

Theorem 6.1: *For every integer $N_1 = q+1-c$ with $|c| \le 2\sqrt{q}$, there exists an elliptic cubic over $GF(q), q = p^h$, with precisely N_1 points, providing one of the following holds:*

(a) $c \not\equiv 0 \pmod{p}$;
(b) $c = 0$ *and h is odd or* $p \not\equiv 1 \pmod 4$;
(c) $c = \pm\sqrt{q}$ *and h is even with* $p \not\equiv 1 \pmod 3$;
(d) $c = \pm 2\sqrt{q}$ *and h is even;*
(e) $c = \pm\sqrt{(2q)}$ *and h is odd with* $p = 2$;
(f) $c = \pm\sqrt{(3q)}$ *and h is odd with* $p = 3$.

Corollary: *The maximum value of $N_q(1)$ of N_1 is given as follows:*

$$N_q(1) = \begin{cases} q + [2\sqrt{q}], & \text{if } h \text{ is odd, } h \ge 3 \text{ and } p|[2\sqrt{q}] \\ q + 1 + [2\sqrt{q}], & \text{otherwise.} \end{cases}$$

For references on the number of isomorphism classes of cubics with a given number of points, see [11]. The smallest case was done in 1915 by Dickson, who classified cubic curves over $GF(2)$. Up to projective equivalence there are 10 irreducible cubics of which 6 are non-singular. Of these, 5 are elliptic with at least one inflexion, and these are the inequivalent ones up to isomorphism. If isomorphism over $\overline{GF(2)}$ is considered, there are only 2 curves, whose affine equations are

$$\begin{aligned} y^2 + y &= x^3, \\ y^2 + xy &= x^3 + x. \end{aligned}$$

See [3], [9].

We now consider the codes $S(D, E)$ on a plane elliptic curve C with N_1 rational points. As before n is the length of the code of dimension k and $r = n - k$.

Theorem 6.2. (Liu-Kumar [19]) *Let $0 < r < N_1$ with $(r!, N_1) = 1$. Then the length n of $S(D,E)$ satisfies*

$$n \leq [\tfrac{1}{2}\{N_1 + (r-1)^2\}] \quad \textit{for } r \textit{ even},$$
$$n \leq [\tfrac{1}{2}\{N_1 + r(r-1)\}] \quad \textit{for } r \textit{ odd}.$$

The points on C form an abelian group in the following way. Take any point O as the identity. The join of any two points P_1 and P_2 meets C again at P_3, and OP_3 meets C again at P_4; define $P_4 = P_1 \oplus P_2$. If $P_1 = P_2$, then P_1P_2 is the tangent at P_1. Other coincidences are treated similarly; see [12] for further details.

Three points P, Q, R of C are collinear if and only if $P \oplus Q \oplus R = O'$, where O' is the point of C at which the tangent at O meets C again.

On the elliptic curve C, consider the code $S(D,E)$ with $D = \Sigma_1^n P_i$, $E = \Sigma m_i Q_i$ and $m = \Sigma m_i$. Then, by the corollary to Theorem 4.1, $d = n - m$ or $n - m + 1$. When $d = n = m + 1$, then $S = S(D,E)$ is an MDS code and its dual $S^\perp$ is also; conversely, if $S^\perp$ is MDS, so is S. Hence S is not MDS if and only if $S^\perp$ is not MDS.

Theorem 6.3: (Driencourt-Michon [3]) *On the elliptic curve C, let $P_E = \oplus m_i Q_i$. Then the code $S(D,E)$ is not MDS if and only if there are m points P_{i_j} such that $P_E = P_{i_1} \oplus \ldots \oplus P_{i_m}$.*

In the example of Section 4, let $O = P_1$ as in Table 1. As Q_1, Q_2, Q_3 are collinear, so $Q_1 \oplus Q_2 \oplus Q_3 = P_6$, because $x + y + z = 0$ is the equation of the tangent to the curve at P_1. Similarly, as P_7, P_8, P_9 are collinear, so $P_7 \oplus P_8 \oplus P_9 = P_6$. Thus $S(D,E)$ is not MDS.

For the generalization of Theorem 6.3 to curves of higher genus, see [30].

7. Decoding of $S^\perp$

Following an idea of Justesen, Skorobogatov and Vladut [24] have given a decoding algorithm for $S^\perp = S(D,E)^\perp$ which is now described.

As in Section 4, let $D = \sum_{i=1}^n P_i$, $E = \sum m_i Q_i$, $m = \sum m_i$, $2g - 2 < m < n$. For f in $L(E)$ and $x = (x_1, x_2, \ldots, x_n) \in GF(q)^n$, let

$$s(x,f) = \sum x_i f(P_i):$$

this is the *syndrome of x with respect to f*.

Let $t > 0$ be such that there exists an effective divisor E' of degree b defined over $GF(q)$ satisfying

$$l(E') > t, \tag{7.1}$$

$$m - b > t + 2g - 2. \tag{7.2}$$

Let $g_1, \ldots, g_\alpha$ be a basis for $L(E')$ and let $h_1, \ldots, h_\beta$ be a basis for $L(E - E')$. Hence $g_i h_j \in L(E)$ for all i and j. Write

$$s_{ij}(x) = s(x, g_i h_j).$$

Consider the system of equations in $Y_1, Y_2, \ldots, Y_\alpha$:

$$\sum_{i=1}^{\alpha} s_{ij}(x) Y_i = 0, \quad j = 1, \ldots, \beta. \tag{7.3}$$

For c in $S^\perp$, let e in $GF(q)^n$ be an error vector of weight $w(e) = t$. Let $\{P_1', \ldots, P_t'\} \subset \{P_1, \ldots, P_n\}$ be the set of points where e has non-zero coordinates. The points $P_1', \ldots, P_t'$ are the *error locators* and the corresponding non-zero coordinates $e_1, \ldots, e_t$ are the *error values*. It will be convenient to write $\mathcal{P} = \{P_1, \ldots, P_n\}, \mathcal{P}' = \{P_1', \ldots, P_t'\}$. Put $x = c + e$.

Lemma 7.1: (a) *If* (7.1) *holds, then* (7.3) *has a non-trivial solution.*
(b) *If* (7.2) *holds, then, for any solution* $(y_1, \ldots, y_\alpha)$ *of* (7.3)*, the function* $g_y = y_1 g_1 + \ldots + y_\alpha g_\alpha$ *in* $L(E')$ *satisfies*

$$g_y(P_i') = 0, \ i = 1, \ldots, t.$$

Proof: (a) $\ell(E' - P_1' \ldots - P_t') \geq \ell(E') - t > 0$ by (7.1). Thus there exists g' in $L(E'), g' \neq 0$, such that $g'(P_i') = 0$, $i = 1, \ldots, t$. So $g' = y_1 g_1 + \ldots + y_\alpha g_\alpha$ for some $y_1, \ldots, y_\alpha$ in $GF(q)$, since $g_1, \ldots, g_\alpha$ is a basis for $L(E')$. However,

$$\begin{aligned}
\sum_{i=1}^{\alpha} s_{ij}(x) y_i &= \sum s_{ij}(e) y_i \\
&= \sum_i \sum_\lambda e_\lambda g_i(P_\lambda') h_j(P_\lambda') y_i \\
&= \sum_\lambda e_\lambda h_j(P_\lambda') \sum_i y_i g_i(P_\lambda') \\
&= \sum_\lambda e_\lambda h_j(P_\lambda') g'(P_\lambda') \\
&= 0.
\end{aligned}$$

(b) By Theorem 3.1 and (7.2),

$$\begin{aligned}
\ell(E - E' - \sum P_i') &= m - b - t - g + 1, \\
\ell(E - E') &= m - b - g + 1.
\end{aligned}$$

Consider a map φ similar to the map ϑ of Section 4, where

$$\varphi : L(E - E') \to GF(q)^t$$

is given by

$$\varphi(f) = (f(P'_1), \ldots, f(P'_t)).$$

Then ker $\varphi = L(E - E' - \sum P'_i)$ and φ is surjective. Fix any P'_j and a function h_j in $L(E - E')$ such that $h_j(P'_j) = 1$ and $h_j(P'_i) = 0$ for $i \neq j$. A suitable linear combination of equations (7.3) is $\sum_i s(x, g_i h_j) Y_i = 0$. If $(y_1, \ldots, y_\alpha)$ is a solution of (7.3), then

$$0 = \sum_i s(x, g_i h_j) y_i = \sum_i \sum_\lambda e_\lambda g_i(P'_\lambda) h_j(P'_\lambda) y_i = e_j g_y(P'_\lambda). \quad \blacksquare$$

Theorem 7.2: *There exists a decoding algorithm for $S^\perp$ that corrects t errors.*

Proof: From Lemma 7.1, there exists g' in $L(E')$ such that $g'(P'_i) = 0$ for each of the locators $P'_1, \ldots, P'_t$ of the error vector e. Suppose $P'_1, \ldots, P'_u$ are the zeros of g' in $\mathcal{P}$. Consider the system

$$\sum_{i=1}^{u} f_j(P'_i) Y_i = s(x, f_j), j = 1, \ldots, \gamma, \tag{7.4}$$

where $f_1, \ldots, f_\gamma$ is a basis for $L(E)$.

Now we show that (7.4) has the unique solution $(e_1, \ldots, e_t, 0, \ldots, 0)$. First, it is a solution by the definition of a syndrome $s(x, f)$. Suppose that $(y_1, \ldots, y_u)$ is a different solution of (7.4). Then $y' = (y_1 - e_1, \ldots, y_t - e_t, y_{t+1}, \ldots, y_u)$ is also a solution, with zero syndromes and thus giving a non-zero codeword of $S^\perp$. Its locators are among $P'_1, \ldots, P'_u$; so $w(y') \leq u$. Since u is the number of zeros of g' in $\mathcal{P}$ and $g' \in L(E')$, so $u \leq \deg E' = b$. However, $b < m - t - 2g + 2 < 2g + 2$ by (7.2). Hence $u < m - 2g + 2 \leq d^\perp$, by Theorem 4.1, Corollary 2. Thus $w(y') < d^\perp$, a contradiction. $\blacksquare$

For further details, see [24].

References:

1. Y. Driencourt and J.F. Michon, Remarques sur les codes géométriques, *C.R. Acad. Paris Sér. I*, 301 (1985), 15-17.
2. Y. Driencourt and J.F. Michon, Rapport sur les codes géométriques, Universités Aix-Marseille II et Paris 7, 1986.

3. Y. Driencourt and J.F Michon, Elliptic codes over fields of characteristic 2, *J. Pure Appl. Algebra* 45 (1987), 15-39.

4. A. Garcia and P. Viana, Weierstrass points on certain non-classical curves, *Arch. Math.* 46 (1986), 315-322.

5. V.D. Goppa, Codes on algebraic curves, *Soviet Math. Doklady 24* (1981), 170-172.

6. V.D. Goppa, Algebraico-geometric codes, *Math. USSR-Isv. 21* (1983), 75-91.

7. V.D. Goppa, Codes and information, *Russian Math Surveys 39* (1984), 87-141.

8. J.P. Hansen, Codes on the Klein quartic, ideals, and decoding, *IEEE Trans. Inform. Theory IT-33* (1987), 923-925.

9. J.W.P. Hirschfeld, *Projective geometries over finite fields*, Oxford, 1979.

10. J.W.P. Hirschfeld, Linear codes and algebraic curves, in *Geometrical combinatorics* (Editors F.C. Holroyd and R.J. Wilson), Pitman, 1984, 35-53.

11. J.W.P. Hirschfeld, *Finite projective spaces of three dimensions*, Oxford, 1985.

12. J.W.P. Hirschfeld and J.A. Thas, Sets with more than one representation as an algebraic curve, *Finite geometries and combinatorial designs*, American Math. Soc., to appear.

13. Y. Ihara, Some remarks on the number of rational points of algebraic curves over finite fields, *J. Fac. Sci. Univ. Tokyo Sect. IA Math. 28* (1981), 721-724.

14. Y. Ihara, On modular curves over finite fields, *Discrete subgroups of Lie groups and applications to modules*, Oxford, 1975, 161-203.

15. G.L. Katsman, M.A. Tsfasman and S.G. Vladut, Modular curves and codes with a polynomial construction, *IEEE Trans. Inform. Theory, IT-30* (1984), 353-355.

16. G. Lachaud, Les codes géométriques de Goppa, *Sém. Bourbaki 37 ème année*, 1984-85, no. 641, *Astérisque 133-134* (1986), 189-207.

17. G. Lachaud, Sommes d'Eisenstein et nombre de points de certaines courbes algébriques sur les corps finis, *C.R. Acad. Sci. Paris Sér. I. 305* (1987), 729-732.

18. G. Lachaud and J. Wolfmann, Sommes de Kloosterman, courbes elliptiques et codes cycliques en caractéristique 2, *C.R. Acad. Sci. Paris Sér. I. 305* (1987), 881-883.

19. C.M. Liu and P.V. Kumar, On the maximum length of MDS Goppa codes on elliptic curves, preprint, 1987.

20. Y.I. Manin, What is the maximal number of points on a curve over $\mathbf{F}_2$, *J. Fac. Sci. Univ. Tokyo Sect. IA Math. 29* (1981), 715-720.

21. J.F. Michon, Amélioration des paramètres des codes géométriques de Goppa, preprint, 1987.

22. J.P. Serre, Sur le nombre de points rationnels d'une courbe algébrique sur un corps fini, *C.R. Acad. Sci. Paris Sér. I 296* (1983), 397-402.

23. J.P. Serre, Nombre de points des courbes algébriques sur $\mathbf{F}_q$, *Séminaire de Théorie des Nombres de Bordeaux*, Expose no. 22, 1983.

24. A.N. Skorobogatov and S.G. Vladut, On the decoding of algebraic-geometric codes. Preprint, 1988.

25. H. Stichtenoth, Which extended Goppa codes are cyclic? Preprint, 1987.

26. H. Stichtenoth, Geometric Goppa codes of genus 0 and their automorphism groups, preprint, 1987.

27. M.A. Tsfasman, Goppa codes that are better than the Varshamov-Gilbert bound, *Problems Inform. Transmission 18* (1982), 163-166.

28. M.A. Tsfasman, S.G. Vladut and I. Zink, Modular curves, Shimura curves and Goppa codes, better than Varshamov-Gilbert bound, *Math. Nachr. 109* (1982), 21-28.

29. J.H. van Lint and T.A. Springer, Generalized Reed-Solomon codes from algebraic geometry, *IEEE Trans. Information Theory IT-33* (1987), 305-309.

30. J.H. van Lint and G. van der Geer, *Introduction to coding theory and algebraic geometry*, Birkhäuser, 1988.

31. A.T. Vasquez, Some explicit self-dual families of error codes based on Goppa's algebraico-geometric construction, preprint, 1987.

32. S.G. Vladut and Y.I. Manin, Codes linéaires et courbes modulaires, *Publications Mathématiques de L'Université Pierre et Marie Curie* no. 72 (translated from the Russian by M. Deza and D. Le Brigand), 1984; English version in J. Soviet Math. 30 (1985), 2611-2643.

33. S.G. Vladut and V.G. Drinfeld, Number of points on an algebraic curve, *Functional Analysis 17* (1983), 53-54.

34. M. Wirtz, On the parameters of Goppa codes, preprint, 1987.

35. M. Wirtz, Verallegemeinerte Goppa-Codes, Diplomarbeit, Universität Munster, 1986.

36. J. Wolfmann, Nombre de points rationnels de courbes algébriques sur des corps finis associées à des codes cycliques, *C.R. Acad. Sci. Paris Sér. I 305* (1987), 345-348.

Solution of a Classical Problem on Finite Inversive Planes

J. A. Thas

Abstract

Let $\mathcal{I}$ be a finite inversive plane of odd order q, $q \notin \{11, 23, 59\}$. If for at least one point p of $\mathcal{I}$ the internal affine plane $\mathcal{I}_p$ is Desarguesian, then $\mathcal{I}$ is Miquelian. Other formulation: for $q \notin \{11, 23, 59\}$ the finite Desarguesian affine plane of odd order q has a unique extension; this extension is the Miquelian inversive plane of order q. As a direct corollary we obtain a computer-free proof of the uniqueness of the inversive plane of order 7.

1. Finite Inversive Planes

A $3-(q^2+1, q+1, 1)$ design is called an *inversive plane* of order q [7,12]. The blocks of an inversive plane are called *circles*. Up to isomorphism, there is a unique inversive plane of order q, with $q \in \{2, 3, 4, 5, 7\}$ [4,8,9,24]. For $q = 7$ this result was obtained by R. H. F. Denniston with the aid of a computer.

Let $\mathcal{I}$ be an inversive plane of order q. For any point p of $\mathcal{I}$, the points of $\mathcal{I}$ different from p, together with the circles containing p (minus p), form a $2-(q^2, q, 1)$ design, i.e., an *affine plane* of order q. That affine plane is denoted by $\mathcal{I}_p$, and is called the *internal plane* of $\mathcal{I}$ at p.

Let C be a circle of the inversive plane $\mathcal{I}$, let $p \in C$ and let p' be a point not on C. Since $\mathcal{I}_p$ is an affine plane it follows that there is a unique circle C' through p and p' which has only p in common with C. We say that the circles C_1 and C_2 are *tangent* if $C_1 = C_2$ or $|C_1 \cap C_2| = 1$. Now it is clear that tangency is an equivalence relation on the set of all circles.

If O is an *ovoid* [7] of $PG(3, q)$, then the points of O together with the intersections $\pi \cap O$, with π a non-tangent plane of O, form an inversive plane of order q. Such an inversive plane is called *egglike*. For an egglike inversive plane $\mathcal{I}$ each internal plane $\mathcal{I}_p$ is Desarguesian. By a celebrated theorem of P. Dembowski [7] each finite inversive plane of even order is egglike. If the ovoid O is an elliptic quadric, then the corresponding inversive plane is called *classical* or *Miquelian*. By a celebrated theorem of A. Barlotti [14] each egglike inversive plane of odd order is Miquelian. Moreover for odd order no other inversive planes are known. The following question is an old, but fundamental, problem: if each internal plane $\mathcal{I}_p$ of the inversive plane $\mathcal{I}$ of odd order q is Desarguesian, prove then that $\mathcal{I}$ is Miquelian. In [23] we gave an answer on that question for $q \equiv 1 \pmod 4$ under the even weaker assumption that $\mathcal{I}_p$ is Desarguesian for at least one point p. In this paper we shall give an answer for all odd q with $q \notin \{11, 23, 59\}$. Finally we remark that the corresponding problem for Laguerre and Minkowski planes of odd order has been solved by Y. Chen and G. Kaerlein in 1973 [5,18].

2. Plane Models of the Classical Circle Geometries

Let O be an elliptic quadric of $PG(3,q)$, let $p \in O$, and let π be a plane of $PG(3,q)$ not containing p. The intersection of π and the tangent plane π_p of O at p is denoted by L. By projection ζ of $O - \{p\}$ from p onto π, the points of $O - \{p\}$ are mapped onto the q^2 points of $\pi - L$, the circles of O through p (minus p) are mapped onto the $q^2 + q$ lines of π different from L, and the circles of O not through p are mapped by ζ onto the $q^3 - q^2$ irreducible conics of π containing two points $r, s \in L$ which are conjugate with respect to the quadratic extension $GF(q^2)$ of $GF(q)$. In this way we obtain the well-known plane model of the classical inversive plane.

Let H be a hyperbolic quadric of $PG(3,q)$, let $p \in H$, and let π be a plane of $PG(3,q)$ not containing p. The intersection of π and the tangent plane π_p of H at p is denoted by L. The lines R and S on H through p intersect L in the respective points $r, s \in L$. By projection ζ of $H - \{p\}$ from p onto π the points of $H - (R \cup S)$ are mapped onto the q^2 points of $\pi - L$, all points of $R - \{p\}$ (respectively, $S - \{p\}$) are mapped onto r (respectively, s), all lines of H not through p are mapped onto the $2q$ lines of π containing just one of r, s, all irreducible conics of H through p (minus p) are mapped onto the $q^2 - q$ lines of π containing neither r nor s, and the irreducible conics of H not through p are mapped by ζ onto the $q^3 - q^2$ irreducible conics of π through r and s. In fact, we described here the well-known plane model of the classical Minkowski plane.

Let K be a quadratic cone with vertex v of $PG(3,q)$, let $p \in K - \{v\}$, and let π be a plane not containing p. The intersection of π and the tangent plane π_p of K at p is denoted by L. The line R on K through p intersects L in the point $r \in L$. By projection ζ of $K - \{p\}$ from p onto π, the points of $K - R$ are mapped onto the q^2 points of $\pi - L$, all points of $R - \{p\}$ are mapped onto r, all lines of K not through p are mapped onto the q lines of π through r, but distinct from L, all irreducible conics of K through p (minus p) are mapped onto the q^2 lines of π not containing r, and the irreducible conics of K not through p are mapped by $\mathcal{I}$ onto the $q^3 - q^2$ irreducible conics of π which are tangent to L at r. In fact, we described here the well-known plane model of the classical Laguerre plane.

3. Flocks

3.1. Flocks of ovoids

Let O be an ovoid of $PG(3,q)$. A partition of all but two points of O into $q-1$ disjoint ovals [7] is called a *flock* of O. The two remaining points are called the *carriers* of the flock. If L is a line of $PG(3,q)$ having no points in common with O, then the $q-1$ planes through L which are non-tangent to O intersect O in the elements of a flock F. Such a flock is called *linear*.

Let F be any flock of an ovoid O of $PG(3,q)$. Then F is necessarily linear. In the odd case this was proved by W. F. Orr [10], in the even case by J. A. Thas [10,19].

3.2. Flocks of quadratic cones

Let K be a quadratic cone with vertex v of $PG(3,q)$. A partition of $K-\{v\}$ into q irreducible conics is called a *flock* of K. If L is a line of $PG(3,q)$ having no points in common with K, then the q planes through L but not through v intersect K in the elements of a flock K. Such a flock is called *linear*.

For $q = 2,3,4$ any flock of the quadratic cone K is linear [22]; for $q = 5,7,8$ all flocks of K were determined by F. De Clerck, H. Gevaert, and J. A. Thas [6]. Examples of nonlinear flocks of quadratic cones exist for any odd prime power q with $q > 3$, and any $q = 2^{2e+1}$ with $e \geq 1$ [11,22].

3.3. Flocks of hyperbolic quadrics

Let H be a hyperbolic quadric of $PG(3,q)$. A partition of H into $q+1$ irreducible conics is called a *flock* of H. If L is a line of $PG(3,q)$ having no points in common with H, then the $q+1$ planes through L intersect H in the elements of a flock F. Such a flock is called *linear*.

Let F be a flock of the hyperbolic quadric H of $PG(3,q)$, with q even. Then it was shown by J. A. Thas [10,20] that F is necessarily linear.

Next, let H be a hyperbolic quadric of $PG(3,q)$, with q odd. In the set of all irreducible conics of H we define the following equivalence relation [10,20,21]: two conics C_1 and C_2 are equivalent if and only if there is an irreducible conic C on H which is tangent to both C_1 and C_2. There are two equivalence classes, denoted by I and II. Let L be a line having no point in common with H, and let L' be the polar line of L with respect to H. The set of all conics of class I (respectively, II) containing L is denoted by V (respectively, V'). For $q \equiv 1 \pmod 4$ the set of all conics of class II (respectively, I) containing L' is denoted by W (respectively, W'); for $q \equiv -1 \pmod 4$ the set of all conics of class I (respectively, II) containing L' is denoted by W (respectively, W'). Then it was shown by J. A. Thas [10,20] that $V \cup W$ (respectively, $V' \cup W'$) is a (non-linear) flock of H. By most authors these flocks are called *Thas flocks*.

In [1] L. Bader shows that for $q = 11, 23, 59$ the hyperbolic quadric H of $PG(3, q)$ has a flock which is neither linear nor a Thas flock. These flocks were independently discovered by N. L. Johnson [15], and for $q = 11, 23$ also by R. D. Baker and G. L. Ebert [3]. Since these flocks are constructed via the exceptional nearfields, L. Bader calls these flocks the *exceptional flocks*.

In [23] J. A. Thas proves that for $q \equiv 1 \pmod 4$ the hyperbolic quadric H of $PG(3, q)$ admits only linear and Thas flocks. Relying on a crucial theorem in [23] and some theorems on Bol planes by M. Kallaher [16,17], L. Bader and G. Lunardon [2] prove that any flock of the hyperbolic quadric H of $PG(3, q)$, q odd, either is a linear flock, or a Thas flock, or an exceptional flock.

4. Two Lemmas

Lemma 1: *Let H be a hyperbolic quadric of $PG(3, q)$, let $p \in H$ and let R and S be the lines on H through p. If the irreducible conics $C_1, C_2, \ldots, C_{q+1}$ on H define a partition of $H - (R \cup S)$ and if exactly one of these conics contains p, then $\{C_1, C_2, \ldots, C_{q+1}\}$ is a flock of H.*

Proof: Suppose that $F = \{C_1, C_2, \ldots, C_{q+1}\}$ is not a flock of H. Then at least one of the points of $(R \cup S) - \{p\}$, say p', is on at least two elements of F, say C_1 and C_2. Let $p' \in S$ and call R' the second line on H through p'. Through each point of $R' - \{p'\}$ there passes a unique element of $F - \{C_1, C_2\}$, and, conversely, each element of $F - \{C_1, C_2\}$ contains at most one point of $R' - \{p'\}$. Hence $q = |R' - \{p'\}| \leq |F - \{C_1, C_2\}| = q - 1$, a contradiction. We conclude that F is a flock of H. ∎

Consider the plane model of H described in Section 2. Suppose that the irreducible conics $C'_1, \ldots, C'_q$ through r, s together with the line $M \neq rs$ define a partition of $PG(2, q) - rs$. Further, let $C_1, C_2, \ldots, C_q, C_{q+1}$ be the irreducible conics on H which correspond with $C'_1, C'_2, \ldots, C'_q, M$. Then from Lemma 1 follows that $F = \{C_1, C_2, \ldots, C_{q+1}\}$ is a flock of H.

Lemma 2: *Let K be a quadratic cone of $PG(3, q)$, with q odd, and let R be a line on K. If the irreducible conics $C_1, C_2, \ldots, C_q$ on K define a partition of $K - R$ and if $C_1, C_2, \ldots, C_{(q+1)/2}$ are elements of a flock F of K, then $F = \{C_1, C_2, \ldots, C_q\}$.*

Proof: Let $C \in F - \{C_1, C_2, \ldots, C_{(q+1)/2}\}$ and suppose that $C \notin \{C_1, C_2, \ldots, C_q\}$. The common point of C and R is denoted by p. Each point of $C - \{p\}$ is contained in just one of the conics $C_{(q+3)/2}, C_{(q+5)/2}, \ldots, C_q$, and, conversely, each of these $(q - 1)/2$ conics contains at most two points of $C - \{p\}$. Hence $q = |C - \{p\}| \leq 2.(q - 1)/2 = q - 1$, a contradiction. We conclude that each conic C of $F - \{C_1, C_2, \ldots, C_{(q+1)/2}\}$ is an element of $\{C_1, C_2, \ldots, C_q\}$, i.e., $F = \{C_1, C_2, \ldots, C_q\}$. ▮

Corollary: *Let K be a quadratic cone of $PG(3,q)$, with q odd, and let R be a line on K. If the irreducible conics $C_1, C_2, \ldots, C_q$ on K define a partition of $K - R$ and if the planes of $C_1, C_2, \ldots, C_{(q+1)/2}$ contain a common line N, then $\{C_1, C_2, \ldots, C_q\}$ is the linear flock defined by N.*

Proof: Immediate from Lemma 2. ▮

Consider the plane model of K described in Section 2. Suppose that the irreducible conics $C'_1, C'_2, \ldots, C'_{q-1}$ are tangent to the line L at r, and that $C'_1, C'_2, \ldots, C'_{q-1}$ together with the line $M \neq L$ define a partition of $PG(2,q) - L$. Suppose moreover that $C'_1, C'_2, \ldots, C'_{(q-1)/2}$ all contain the points a', a'' ($a' \neq a''$) of M which are conjugate with respect to the quadratic extension $GF(q^2)$ of $GF(q)$. Further, let $C_1, C_2, \ldots, C_{q-1}, C_q$ be the irreducible conics on K which correspond with $C'_1, C'_2, \ldots, C'_{q-1}, M$. Then from the corollary of Lemma 2 follows that $F = \{C_1, C_2, \ldots, C_q\}$ is a linear flock, so that $C'_1, C'_2, \ldots, C'_{q-1}$ all contain a', a''.

5. Main Theorem

Theorem: *Let $\mathcal{I}$ be an inversive plane of odd order q, $q \notin \{11, 23, 59\}$. If for at least one point p of $\mathcal{I}$ the internal plane $\mathcal{I}_p$ is Desarguesian, then $\mathcal{I}$ is Miquelian.*

Proof: Since there is a unique inversive plane of order 3 and 5, we may assume that $q \geq 7$. The proof will be subdivided into six parts.

Part 1. Let $\mathcal{I}$ be an inversive plane of odd order q, $q \notin \{11, 23, 59\}$. Suppose that there is at least one point p of $\mathcal{I}$ for which the internal plane $\mathcal{I}_p$ is the Desarguesian affine plane $AG(2,q)$. The line at infinity of $\mathcal{I}_p = AG(2,q)$ is denoted by L, and the projective plane defined by $AG(2,q)$ is $PG(2,q)$. The circles of $\mathcal{I}$ through p (minus p) are the lines of $AG(2,q)$. The $q^3 - q^2$ circles not containing p are $(q+1)$-arcs [13] of $PG(2,q)$. By a celebrated theorem of B. Segre [13], any $(q+1)$-arc of $PG(2,q)$, with q odd, is an irreducible conic. Hence the $q^3 - q^2$ circles not containing p are irreducible conics of $PG(2,q)$. Each of these conics intersects L in distinct points which are conjugate with respect to the quadratic extension $GF(q^2)$ of $GF(q)$.

Let $r, s \in \mathcal{I} - \{p\}$ and consider the $q+1$ circles of $\mathcal{I}$ through r and s. In the plane $AG(2,q)$ these circles are the line rs and q irreducible conics $C_1, C_2, \ldots, C_q$ through r and s. Now we consider the plane model of the hyperbolic quadric H of $PG(3,q)$, defined by the points r and s of $PG(2,q)$. By Lemma 1, the line L together with the conics $C_1, C_2, \ldots, C_q$ define a flock F of H. By 3.3 the flock F is linear or a Thas flock.

First suppose that F is linear. In such a case we have $C_1 \cap L = C_2 \cap L = \ldots = C_q \cap L = \{\overline{r}, \overline{s}\}$. The points $\overline{r}$ and $\overline{s}$ will be called the *carriers* of the flock F or the *carriers* defined by r and s.

Next, assume that F is a Thas flock. In such a case $(q-1)/2$ of the conics $C_1, C_2, \ldots, C_q$ contain two common points $\overline{r}$ and $\overline{s}$ of L, say $C_1 \cap L = C_2 \cap L = \ldots = C_{(q-1)/2} \cap L = \{\overline{r}, \overline{s}\}$. Let $r\overline{r} \cap s\overline{s} = \{\overline{\overline{r}}\}$ and $r\overline{s} \cap \overline{r}s = \{\overline{\overline{s}}\}$. Then the conics $C_{(q+1)/2}, \ldots, C_q$ contain the points $\overline{\overline{r}}$ and $\overline{\overline{s}}$. Clearly $\overline{\overline{r}}$ and $\overline{\overline{s}}$ are conjugate with respect to the quadratic extension $GF(q^2)$ of $GF(q)$. Let $C_i \cap L = \{\overline{r}_i, \overline{s}_i\}$, $i = (q+1)/2, (q+3)/2, \ldots, q$. The bundle consisting of all conics of $PG(2, q^2) \supset PG(2, q)$ through $r, s, \overline{\overline{r}}, \overline{\overline{s}}$ intersects L in an involution with fixed points $\overline{r}$ and $\overline{s}$. It follows that $(\overline{r}\,\overline{s}\,\overline{r}_i\,\overline{s}_i) = -1$, with $i = (q+1)/2, (q+3)/2, \ldots, q$. Hence $\{\overline{r}_{(q+1)/2}, \overline{s}_{(q+1)/2}\}, \ldots, \{\overline{r}_q, \overline{s}_q\}$ are nothing else than the $(q+1)/2$ pairs of conjugate points on L (with respect to the quadratic extension of $GF(q)$) which are harmonic conjugate to $\overline{r}$ and $\overline{s}$. Also in this case the points $\overline{r}$ and $\overline{s}$ will be called the *carriers* of the flock F or the *carriers* defined by r and s.

We now fix a point $o \in AG(2, q)$. If $a \in AG(2, q) - \{o\}$, then the carriers defined by o and a will be denoted by a' and a''.

Part 2. *Assume that a and b are distinct points of $AG(2, q) - \{o\}$, with $\{o, a\}$ and $\{o, b\}$ defining linear flocks. Assume moreover that $\{a', a''\} \neq \{b', b''\}$.*

If o, a, b are not collinear, then the circle oab contains the carriers a', a'' but also the carriers b', b'', a contradiction. Hence, o, a, b are collinear in $PG(2, q)$. Let c be a point of the plane $AG(2, q)$ which is not on the line ab. Then the circle oac contains the carriers a', a'', and the circle obc contains the carriers b', b''. It follows that $\{o, c\}$ defines a Thas flock.

(1) $(a'a''b'b'') \neq -1$.

Let c be a point of the plane $AG(2, q)$ which is not on the line ab. The circle oac contains the carriers a', a'', and the circle obc contains the carriers b', b''. If a', a'' are the carriers defined by $\{o, c\}$, then $(a'a''b'b'') = -1$, a contradiction. Analogously, the carriers defined by $\{o, c\}$ are not b', b''. Let c', c'' be the carriers defined by $\{o, c\}$. Then $\{c', c''\}$ is the unique pair for which $(a'a''c'c'') = -1$ and $(b'b''c'c'') = -1$.

Let $c = c_1, c_2, \ldots, c_{q-1}$ be the $q-1$ affine points of $M - \{o\}$, with $M = oc$. The carriers defined by $\{o, c_i\}$ are c' and c'', $i = 1, 2, \ldots, q-1$. For each pair $\{d', d''\}$ of conjugate points on L with $(c'c''d'd'') = -1$, there is a unique circle through o, c_i, d', d'', with $i = 1, 2, \ldots, q-1$. Let C_i be the circle through o, c_i, d', d'', and let C_j be the circle through o, c_j, d', d'', with $i \neq j$. If d is a point of $AG(2, q) - ab$ on C_i and C_j, then d' and d'' are the carriers defined by $\{o, d\}$, a contradiction. Hence C_i and C_j are tangent at o, or the common points of C_i and C_j are o, d', d'' and a point of $AG(2, q)$ on $ab - \{o\}$. If for some pair $\{i, j\}$, with $i, j \in \{1, 2, \ldots, q-1\}$, the circles C_i and C_j contain a common point d of $AG(2, q)$ on $ab - \{o\}$, then it is easy to see that $C_1, C_2, \ldots, C_{q-1}$ all contain d. Hence $C_1, C_2, \ldots, C_{q-1}$ are mutually tangent at o, or $C_1, C_2, \ldots, C_{q-1}$ all contain a common point d of $AG(2,q)$ on $ab - \{o\}$. It is clear that if $\{d', d''\} = \{a', a''\}$ (respectively, $\{d', d''\} = \{b', b''\}$), then $C_1, C_2, \ldots, C_{q-1}$ all contain a (respectively, b).

If $C_1, C_2, \ldots, C_{q-1}$ all contain a common point d of $AG(2,q)$ on $ab - \{o\}$, then the flock defined by $\{o, d\}$ is linear with carriers d' and d''.

Now suppose that $C_1, C_2, \ldots, C_{q-1}$ are mutually tangent at o. By way of contradiction assume that the common tangent of $C_1, C_2, \ldots, C_{q-1}$ at o is not the line ab. Then the circles $C_1, C_2, \ldots, C_{q-1}$ intersect the affine line ab in o and $q-1$ distinct affine points of $ab - \{o, a, b\}$, a contradiction. It follows that ab is the common tangent of $C_1, C_2, \ldots, C_{q-1}$ at o. Hence the set $\{C_1, C_2, \ldots, C_{q-1}\}$ is uniquely defined by the line ab, so for at most one pair $\{d', d''\}$ the circles $C_1, C_2, \ldots, C_{q-1}$ are mutually tangent at o.

Consequently the $(q+1)/2$ pairs $\{d', d''\}$ define either $(q+1)/2$ or $(q-1)/2$ points of $AG(2,q)$ on $ab - \{o\}$. The set of these points is denoted by D. Let c_i be an affine point of $oc - \{o\}$ and let $oc'' \cap c_i c' = \{\bar{c}_i\}$ and $oc' \cap c_i c'' = \{\bar{\bar{c}}_i\}$. Let the conic through $o, c_i, \bar{c}_i, \bar{\bar{c}}_i, d', d''$ be denoted by C_i. Then $C_1, C_2, \ldots, C_{q-1}$ either are tangent to oa at o, or contain $d \in D$. It follows that in $PG(2,q)$ the points o, c', c'' together with the line oc define uniquely the line oa. In fact, in $PG(2,q)$, any line through o, but distinct from oa, together with o, c', c'' define uniquely the line oa. Let M and M' be distinct lines through o, with $M \neq oa \neq M'$. If ζ is any projectivity of $PG(2,q)$ for which $\{c', c''\}^\zeta = \{c', c''\}$, $o^\zeta = o$, and $M^\zeta = M'$, then necessarily $(oa)^\zeta = oa$. This clearly yields a contradiction.

(2) $\quad (a'a''b'b'') = -1.$

Let c be a point of the plane $AG(2,q)$ which is not on the line ab. Then $\{o, c\}$ defines a Thas flock with carriers c' and c''. By way of contradiction assume that $\{a', a''\} \neq \{c', c''\} \neq \{b', b''\}$. The circle oac contains a', a'' and the circle obc contains b', b'', so $(a'a''c'c'') = (b'b''c'c'') = -1$. Let $(c'c''d'd'') = (c'c''e'e'') = -1$, with $\{d', d''\} \neq \{e', e''\}$, $\{a', a''\} \neq \{d', d''\} \neq \{b', b''\}$, $\{a', a''\} \neq \{e', e''\} \neq \{b', b''\}$ (such is possible since $q \geq 7$). Let C be the circle through o, c, d', d'' and let C' be the circle through o, c, e', e''. For any point s on $C - ab$, with $s \neq c$, the circle soc contains d', d'', the circle osa contains a', a'', and the circle osb contains b', b''. Since $(a'a''d'd'') \neq -1$, $(b'b''d'd'') \neq -1$, $(a'a''c'c'') = (b'b''c'c'') = (d'd''c'c'') = -1$, the points c' and c'' are the carriers of $\{o, s\}$. Hence for any point s on $C - ab$ the pair $\{o, s\}$ has carriers c', c''. Analogously, for any point r on $C' - ab$ the pair $\{o, r\}$ has carriers c', c''. Let $r \in C' - ab$, with $r \neq c$, and let C''' be the circle containing o, r, d', d''. Clearly $C \neq C'''$. If C and C''' contain a common point $t \neq o$ not on ab, then the pair $\{o, t\}$ has carriers d', d'', a contradiction. Hence C and C''' either are tangent at o, or contain a common point h on $ab - \{o\}$. Interchanging roles of r and c we see that for each point $u \in C''' - ab$, the pair $\{o, u\}$ has carriers c', c''. Hence for two distinct points r on $C' - ab$, $r \neq c$, the corresponding circles C'''

either are tangent at o, or contain a common point on $ab - \{o\}$. There are at least $q-2$ points r, so there arise at least $q-2$ circles C'''', say $C_1'''', C_2'''', \ldots, C_{q-2}''''$. If two of the circles $C, C_1'''', \ldots, C_{q-2}''''$ contain a common point h on $ab - \{o\}$, then it is easy to see that $C, C_1'''', \ldots, C_{q-2}''''$ all contain h. Hence $C, C_1'''', \ldots, C_{q-2}''''$ are mutually tangent at o, or $C, C_1'''', \ldots, C_{q-2}''''$ all contain a common point h of $AG(2,q)$ on $ab - \{o\}$.

First, assume that $C, C_1'''', \ldots, C_{q-2}''''$ all contain a common point h on $ab - \{o\}$. All these circles contain d' and d''. Hence $\{o, h\}$ defines a linear flock with carriers d', d''. As $(a'a''d'd'') \neq -1$, we have a contradiction by Case (1).

It follows that $C, C_1'''', \ldots, C_{q-2}''''$ are mutually tangent at o. All these circles contain d' and d''. The common tangent at o will be denoted by N. The line L together with the $q-1$ conics $C, C_1'''', \ldots, C_{q-2}''''$ define a partition of $PG(2,q) - N$. If $N \neq ab$, then one of the $q-1$ conics contains b, so also b', b'', a contradiction. Hence $N = ab$. Consequently the $q-1$ circles tangent to ab at o all contain d' and d''.

Analogously, the $q-1$ circles tangent to ab at o all contain e' and e''. This contradiction shows that either $\{c', c''\} = \{a', a''\}$, or $\{c', c''\} = \{b', b''\}$.

The affine points of the line $M = oc$ are denoted by $o, c = c_1, c_2, \ldots, c_{q-1}$. By way of contradiction assume that a', a'' are the carriers defined by at least $(q+1)/2$ pairs $\{o, c_i\}$, say by $\{o, c_1\}, \{o, c_2\}, \ldots, \{o, c_{(q+1)/2}\}$. Let $\{d', d''\}$ be a pair of conjugate points on L with $(a'a''d'd'') = -1$, $\{b', b''\} \neq \{d', d''\}$ and $(b'b''d'd'') \neq -1$. Then there is a unique circle C_i through o, c_i, d', d'', $i = 1, 2, \ldots, (q+1)/2$. If C_i and C_j, with $i \neq j$, contain a common point $t \neq o$ not on ab, then the pair $\{o, t\}$ has carriers d', d'', a contradiction. Hence C_i and C_j, $i \neq j$, either are tangent at o, or contain a common point d on $ab - \{o\}$. If for some pair $\{i, j\}$, with $i, j \in \{1, 2, \ldots, (q+1)/2\}$, the circles C_i and C_j contain a common point d of $AG(2,q)$ on $ab - \{o\}$, then it is easy to see that $C_1, C_2, \ldots, C_{(q+1)/2}$ all contain d. Hence $C_1, C_2, \ldots, C_{(q+1)/2}$ are mutually tangent at o, or $C_1, C_2, \ldots, C_{(q+1)/2}$ all contain a common point d of $AG(2,q)$ on $ab - \{o\}$.

Suppose that $C_1, C_2, \ldots, C_{(q+1)/2}$ all contain a common point d of $AG(2,q)$ on $ab - \{o\}$. In such a case $\{o, d\}$ defines a linear flock with carriers d' and d''. Since $(b'b''d'd'') \neq -1$ we have a contradiction by Case (1).

Hence $C_1, C_2, \ldots, C_{(q+1)/2}$ are mutually tangent at o. The common tangent will be denoted by N. Let $C'_{(q+3)/2}, \ldots, C'_{q-1}$ be the remaining circles which are tangent to N at o. Now we consider the plane model of the quadratic cone K of $PG(3,q)$, defined by the line N and the point $o \in N$ of $PG(2,q)$. The line L together with the conics $C_1, \ldots, C_{(q+1)/2}, C'_{(q+3)/2}, \ldots, C'_{q-1}$ define a partition of $PG(2,q) - N$. The conics $C_1, C_2, \ldots, C_{(q+1)/2}$ all contain d' and d''. Now by the corollary of Lemma 2 the line L together with the conics $C_1, \ldots, C_{(q+1)/2}, C'_{(q+3)/2}, \ldots, C'_{q-1}$ define a linear flock F of K. If $N \neq ab$, then one of the $q-1$ conics contains a, d', d'', a contradiction. Hence $N = ab$. Now, let $\{\overline{d}', \overline{d}''\}$ be a pair of conjugate points on L with $(a'a''\overline{d}'\overline{d}'') = -1$, $\{b', b''\} \neq \{\overline{d}', \overline{d}''\} \neq \{d', d''\}$ and $(b'b''\overline{d}'\overline{d}'') \neq -1$ (such is possible since $q \geq 7$). Again we find $q-1$ circles $\overline{C}_1, \ldots, \overline{C}_{(q+1)/2}, \overline{C}'_{(q+3)/2}, \ldots, \overline{C}'_{q-1}$ tangent to ab at o. Hence $\{C_1, \ldots, C_{(q+1)/2}, C'_{(q+3)/2}, \ldots, C'_{q-1}\} = \{\overline{C}_1, \ldots, \overline{C}_{(q+1)/2}, \overline{C}'_{(q+3)/2}, \ldots, \overline{C}'_{q-1}\}$. It follows that $C_1, \ldots, C'_{q-1}$ all contain d', d'' and $\overline{d}', \overline{d}''$, a contradiction.

Consequently exactly $(q-1)/2$ pairs $\{o, c_i\}$ have a', a'' as carriers, and exactly $(q-1)/2$ pairs $\{o, c_j\}$ have b', b'' as carriers. Let d', d'' be conjugate points on L with $(a'a''d'd'') = -1$, $\{d', d''\} \neq \{b', b''\}$ and $(b'b''d'd'') \neq -1$. Suppose that a', a'' are the carriers defined by $\{o, c_1\}, \{o, c_2\}, \ldots, \{o, c_{(q-1)/2}\}$. Let C_i be the circle through o, c_i, d', d'', with $i = 1, 2, \ldots, (q-1)/2$. Again $C_1, C_2, \ldots, C_{(q-1)/2}$ are mutually tangent at o, or $C_1, C_2, \ldots, C_{(q-1)/2}$ all contain a common point d of $AG(2,q)$ on $ab - \{o, a, b\}$.

First, suppose that $C_1, C_2, \ldots, C_{(q-1)/2}$ are mutually tangent at o. The common tangent will be denoted by N. Let $C'_{(q+1)/2}, C'_{(q+3)/2}, \ldots, C'_{q-1}$ be the remaining circles which are tangent to N at o. Now we consider the plane model of the quadratic cone K of $PG(3,q)$, defined by the line N and the point $o \in N$ of $PG(2,q)$. The circles $C_1, C_2, \ldots, C_{(q-1)/2}$ all contain the points d', d'' of L, so, by the corollary of Lemma 2, the line L together with the conics $C_1, \ldots, C_{(q-1)/2}, C'_{(q+1)/2}, \ldots, C'_{q-1}$ define a linear flock F of K. If $N \neq ab$, then one of the $q-1$ conics contains a, d', d'', a contradiction. Hence $N = ab$. One of the conics $C'_{(q+1)/2}, C'_{(q+3)/2}, \ldots, C'_{q-1}$, say $C'_{(q+1)/2}$, contains $c_{(q+1)/2}$. The carriers defined by $\{o, c_{(q+1)/2}\}$ are b', b'', and $C'_{(q+1)/2}$ contains d', d''. Hence $(b'b''d'd'') = -1$, a contradiction.

Consequently $C_1, C_2, \ldots, C_{(q-1)/2}$ all contain a common point d of $AG(2,q)$ on $ab - \{o, a, b\}$. First, suppose that $\{o, d\}$ defines a linear flock. Since the circle $odc_{(q+1)/2}$ contains d', d'' and $\{o, c_{(q+1)/2}\}$ has carriers b', b'', we have $(b'b''d'd'') = -1$, a contradiction. Hence $\{o, d\}$ defines a Thas flock with carriers d', d''. The circles $odc_i = C_i$, $i = 1, 2, \ldots, (q-1)/2$, all contain d', d'', so $odc_{(q+1)/2}$ intersects L' in points e', e'' with $(d'd''e'e'') = -1$. Since $\{o, c_{(q+1)/2}\}$ has carriers b', b'' and $(b'b''d'd'') \neq -1$, we have $(b'b''e'e'') = -1$. From $(d'd''e'e'') = (b'b''e'e'') = -1$ follows that $\{e', e''\} = \{a', a''\}$. Analogously the circle $odc_{(q+3)/2}$ contains a', a''. Consequently $\{o, d\}$ has carriers a', a'', clearly a contradiction.

Conclusion of this Part 2. All pairs $\{o, a\}$ defining a linear flock have the same carriers a', a'' on L.

Part 3. *Assume that a and b are distinct points of $AG(2,q) - \{o\}$, $o \notin ab$, with $\{o, a\}$ defining a linear flock and with $\{o, b\}$ defining a Thas flock.*

The circle oab intersects L in a' and a''. If b', b'' are the carriers of

$\{o,b\}$, then either $\{a',a''\} = \{b',b''\}$, or $\{a',a''\} \neq \{b',b''\}$ with $(a'a''b'b'') = -1$. Assume that $\{a',a''\} \neq \{b',b''\}$.

Let c be a point of $AG(2,q)$, with $c \notin oa$, $c \notin ob$, and let c', c'' be the carriers of $\{o,c\}$. Suppose that $\{c',c''\} \neq \{a',a''\}$, $\{c',c''\} \neq \{b',b''\}$, and $(b'b''c'c'') \neq -1$. By Part 2 the pair $\{o,c\}$ defines a Thas flock. Considering the circle oac, we see that $(a'a''c'c'') = -1$. Since $(b'b''c'c'') \neq -1$ the circle obc does neither contain c', c'', nor b', b''. If $obc \cap L = \{d',d''\}$, then $(c'c''d'd'') = (b'b''d'd'') = -1$, so $\{a',a''\} = \{d',d''\}$. There is a unique circle through o, b, a', a'', namely the circle oab. Since obc contains o, b, a', a'', we have $c \in oab$. Consider a line T through o, with $T \notin \{oa, ob, oc\}$, and let t be an affine point of T not on the circle oab. If the carriers of $\{o,t\}$ are c', c'', then, interchanging roles of t and c, we see that $t \in oab$, a contradiction. Suppose that $\{t',t''\} \neq \{a',a''\}$. Considering the circle oat we obtain $(a'a''t't'') = -1$. The circle oat is the unique circle through o and t which contains a', a''. Since $oat \neq oct$, the circle oct does not contain a', a''. If oct contains neither c', c'' nor t', t'', then $L \cap oct = \{u',u''\}$ with $(c'c''u'u'') = (t't''u'u'') = -1$, so $\{u',u''\} = \{a',a''\}$, a contradiction. It follows that oct contains either c', c'' or t', t''. Consequently $(c'c''t't'') = -1$. Interchanging roles of b and c, we obtain $(b'b''t't'') = -1$. From $(c'c''t't'') = (b'b''t't'') = -1$ it follows that $\{t',t''\} = \{a',a''\}$, a contradiction. So we conclude that a', a'' are the carriers of $\{o,t\}$. First, suppose that the flock defined by $\{o,t\}$ is linear. Then the circle otb contains a', a''. Since oab is the unique circle through o, b, a', a'', we necessarily have $t \in oab$, a contradiction. So the flock defined by $\{o,t\}$ is always a Thas flock. Now let d', d'' be conjugate points on L with $(b'b''d'd'') = -1$, $\{d',d''\} \neq \{a',a''\}$ and $(a'a''d'd'') \neq -1$. Let C be the circle through o, b, d', d''. The pair $\{d',d''\}$ can always be chosen in such a way that C intersects T in two distinct points o and u. Since $u \notin oab$, the pair $\{o,u\}$ has carriers a', a''. It follows that $(a'a''d'd'') = -1$, a final contradiction.

Next, let c be a point of $AG(2,q)$, with $c \notin oa$, $c \notin ob$, and let c', c'' be the carriers of $\{o,c\}$. Suppose that $\{c',c''\} \neq \{a',a''\}$, $\{c',c''\} \neq \{b',b''\}$ and $(b'b''c'c'') = -1$. By Part 2 the pair $\{o,c\}$ defines a Thas flock. Considering the circle oac, we see that $(a'a''c'c'') = -1$. Choose conjugate points d', d'' on L with $(c'c''d'd'') = -1$ and $\{a',a''\} \neq \{d',d''\} \neq \{b',b''\}$. Next, choose conjugate points e', e'' on L with $(c'c''e'e'') = -1$, $\{a',a''\} \neq \{e',e''\} \neq \{b',b''\}$ and $\{d',d''\} \neq \{e',e''\}$ (such is possible since $q \geq 7$). Then $(a'a''d'd'') \neq -1$ and $(b'b''d'd'') \neq -1$. Let C be the circle through o, c, d', d''. For any point s on $C - oa$, with $s \neq c$, the circle soc contains d', d'' and the circle osa contains a', a''. Since $(a'a''d'd'') \neq -1$ and $(d'd''c'c'') = (a'a''c'c'') = -1$, the pair $\{o,s\}$ has carriers c', c''. Hence for any point s on $C - oa$ the pair $\{o,s\}$ has carriers c', c''. Let $s \in C - oa$ and let C' be the circle through o, s, e', e''. For any point f on $C' - oa$, with $f \neq s$, the circle osf contains e', e'' and the circle ofa contains a', a''. Since $(a'a''e'e'') \neq -1$ and $(e'e''c'c'') = (a'a''c'c'') = -1$, the pair $\{o,f\}$

has carriers c', c''. On $C - oa$ there are at least $q - 1$ points s, so there arise at least $q-1$ circles C', say $C'_1, C'_2, \ldots, C'_{q-1}$. For two distinct points s on $C - oa$ the corresponding circles C' either are tangent at o, or contain a common point h on $oa - \{o\}$. If two of the circles C' contain a common point h on $oa - \{o\}$, then it is easy to see that all circles C' contain h. Hence $C'_1, C'_2, \ldots, C'_{q-1}$ are mutually tangent at o, or $C'_1, C'_2, \ldots, C'_{q-1}$ all contain a common point h of $AG(2,q)$ on $oa - \{o\}$. First, assume that the circles $C'_1, C'_2, \ldots, C'_{q-1}$ all contain a common point h on $oa - \{o\}$. Then $\{o, h\}$ defines a linear flock with carriers e', e''. This contradicts Part 2. Hence $C'_1, C'_2, \ldots, C'_{q-1}$ are mutually tangent at o. The common tangent at o will be denoted by N. The line L together with the $q-1$ conics $C'_1, C'_2, \ldots, C'_{q-1}$ define a partition of $PG(2,q) - N$. If $N \neq oa$, then one of the $q-1$ conics contains a, so also a', a'', clearly a contradiction. Consequently $N = oa$. We conclude that the $q-1$ circles tangent to oa at o all contain e', e''. Analogously, the $q-1$ circles tangent to oa at o all contain d', d'', a contradiction.

It follows that for each point c of $AG(2,q)$, with $c \notin oa$ and $c \notin ob$, the pair $\{o, c\}$ has carriers $\{a', a''\}$ or $\{b', b''\}$.

Let $(b'b''d'd'') = (b'b''e'e'') = -1$, $(a'a''d'd'') \neq -1$, $(a'a''e'e'') \neq -1$, $\{d', d''\} \neq \{a', a''\} \neq \{e', e''\}$, $\{d', d''\} \neq \{e', e''\}$ (such is possible since $q \geq 7$). Further, let C be the circle through o, b, d', d'' and let C' be the circle through o, b, e', e''. For any point $r \in C - oa$, with $r \neq b$, the circle oar contains a', a'' and the circle orb contains d', d''. Since $(a'a''d'd'') \neq -1$ and $(a'a''b'b'') = (d'd''b'b'') = -1$, the pair $\{o, r\}$ has carriers b', b''. Consequently, for any point $r \in C - oa$ the pair $\{o, r\}$ has carriers b', b''. Analogously, for any point $s \in C' - oa$, the pair $\{o, s\}$ has carriers b', b''. Let $r \in C - oa$ and let C'' be the circle through o, r, e', e''. On $C - oa$ there are at least $q-1$ points r, so there arise at least $q-1$ circles C'', say $C' = C''_1, C''_2, \ldots, C''_{q-1}$. Interchanging roles of b and r, we see that for any point f on $C''_i - oa$ the pair $\{o, f\}$ has carriers b', b''. Such as in the preceding sections it follows that $C''_1, C''_2, \ldots, C''_{q-1}$ are mutually tangent at o, or that $C''_1, C''_2, \ldots, C''_{q-1}$ all contain a common point h of $AG(2,q)$ on $oa - \{o\}$. First, assume that the circles $C''_1, C''_2, \ldots, C''_{q-1}$ all contain a common point h on $oa - \{o\}$. Then $\{o, h\}$ defines a linear flock with carriers e', e''. This contradicts Part 2. Hence $C''_1, C''_2, \ldots, C''_{q-1}$ are mutually tangent at o. The common tangent at o will be denoted by N. The line L together with the $q-1$ conics $C''_1, C''_2, \ldots, C''_{q-1}$ define a partition of $PG(2,q) - N$. If $N \neq oa$, then one of the $q-1$ conics contains a, so also a', a'', clearly a contradiction. Consequently $N = oa$. We conclude that the $q-1$ circles tangent to oa at o all contain e', e''. Analogously, the $q-1$ circles tangent to oa at o all contain d', d'', a final contradiction.

Conclusion of this Part 3. If o, a, b are non-collinear points of $AG(2,q)$, with $\{o, a\}$ defining a linear flock and $\{o, b\}$ defining a Thas flock, then $\{a', a''\} = \{b', b''\}$.

Part 4. *Assume that each point of $AG(2,q)-\{o\}$ defines a Thas flock.*

Suppose we can find non-collinear points o, a, b in $AG(2,q)$, with $\{a', a''\} \neq \{b', b''\}$ and $(a'a''b'b'') \neq -1$. If d', d'' are the common points of L and the circle oab, then $\{d', d''\}$ is the unique pair on L defined by $(a'a''d'd'') = (b'b''d'd'') = -1$. Let D be one of the circles through o, b, b', b''. Intersecting D with the $(q-1)/2$ circles through o, a, a', a'', we find at least $(q-3)/2$ (≥ 2) affine points $d_1, d_2, \ldots, d_{(q-3)/2}$ on D. Clearly $b \neq d_i \neq a$, $i = 1, 2, \ldots, (q-3)/2$. Considering the circles od_ia and od_ib, we see that d', d'' are the carriers of the pair $\{o, d_i\}$, $i = 1, 2, \ldots, (q-3)/2$. Let c', c'' be conjugate on L with $(b'b''c'c'') = -1$, $(c'c''d'd'') \neq -1$ and $\{c', c''\} \neq \{d', d''\}$. Further, let C be the circle through o, b, c', c''. Intersecting C with the $(q-1)/2$ circles through o, d_i, d', d'', we find at least $(q-3)/2$ points $p_1^i, p_2^i, \ldots p_{(q-3)/2}^i$ on C, $i = 1, 2, \ldots, (q-3)/2$. Since $(c'c''d'd'') \neq -1$ and $(b'b''c'c'') = (b'b''d'd'') = -1$, each pair $\{o, p_j^i\}$ has carriers b', b''. If $p_j^i = p_{j'}^{i'}$, for $i \neq i'$, then $\{o, p_j^i\}$ has carriers d', d'', clearly a contradiction. So on $C - \{o\}$ there arise at least $\frac{(q-3)^2}{4} \geq q-3$ points p_j^i, with $\{o, p_j^i\}$ having as carriers b', b''. At least $q-4$ of these points are not on the line oa. For these $q-4$ points the circle oap_j^i contains d', d'' since $(a'a''d'd'') = (b'b''d'd'') = -1$ and $(a'a''b'b'') \neq -1$. Hence d', d'' are the carriers of $\{o, a\}$, a contradiction.

Next, suppose we can find non-collinear points o, a, b in $AG(2,q)$, with $\{a', a''\} \neq \{b', b''\}$ and $(a'a''b'b'') = -1$. By the previous paragraph, for each point c not on oa or ob the pair $\{o, c\}$ has carriers a', a'', or b', b'', or c', c'' with $(a'a''c'c'') = (b'b''c'c'') = -1$. Let t', t'' be conjugate points on L for which $(a'a''t't'') = -1$ and $\{b', b''\} \neq \{t', t''\} \neq \{c', c''\}$. Further, let C be the circle containing o, a, t', t''. Since $(b'b''t't'') \neq -1$ and $(c'c''t't'') \neq -1$, for each point $r \in C - ob$ the pair $\{o, r\}$ has carriers a', a''. Now let u', u'' be conjugate points on L for which $(b'b''u'u'') = -1$ and $\{a', a''\} \neq \{u', u''\} \neq \{c', c''\}$. Further, let C' be the circle containing o, b, u', u''. Since $(a'a''u'u'') \neq -1$ and $(c'c''u'u'') \neq -1$, for each point $s \in C' - ob$ the pair $\{o, s\}$ has carriers b', b''. Clearly $\{u', u''\} \neq \{t', t''\}$, $C \neq C'$, and the circle oab intersects L in either a', a'', or b', b'', or c', c''. Hence also $C \neq oab \neq C'$. If $C \cap C' = \{o, f\}$, then $f \notin oa$ and $f \notin ob$, so $\{o, f\}$ has carriers $\{a', a''\}$ but also $\{b', b''\}$, a contradiction. It follows that C and C' are mutually tangent at o. Since there are $(q-3)/2$ (≥ 2) pairs $\{u', u''\}$ there arise $(q-3)/2$ circles C', say $C_1', C_2', \ldots, C_{(q-3)/2}'$. Each C_i' is tangent to C at o, so $C_1', C_2', \ldots, C_{(q-3)/2}'$ are mutually tangent at o. Since $C_1', C_2', \ldots, C_{(q-3)/2}'$ all contain b, we have a contradiction.

Conclusion of this Part 4. At least one point of $AG(2,q) - \{o\}$ defines a linear flock.

Part 5. *Assume that a and b are distinct points of $AG(2,q) - \{o\}$, with $\{o, a\}$ defining a linear flock and with $\{o, b\}$ defining a Thas flock.*

First, suppose that o, a, b are not collinear. By the conclusion of Part 3 we have $\{a', a''\} = \{b', b''\}$. Assume that for each affine point of $ob - \{o\}$ the corresponding flock is of Thas type. Let d', d'' be conjugate on L with $(a'a''d'd'') = -1$. For each $b_i \in ob - \{o\}$, with $i = 1, 2, \dots, q-1$, there is a unique circle C_i through o, b_i, d', d''. By the conclusions of Parts 2 and 3, either $C_1, C_2, \dots, C_{q-1}$ are mutually tangent at o, or $C_1, C_2, \dots, C_{q-1}$ all contain a common affine point on $oa - \{o, a\}$. If $C_1, C_2, \dots, C_{q-1}$ all contain a common affine point r on $oa - \{o, a\}$, then $\{o, r\}$ defines a linear flock with carriers $\{d', d''\}$, contradicting the conclusion of Part 2. So $C_1, C_2, \dots, C_{q-1}$ are mutually tangent at o. The common tangent line at o will be denoted by N. If $N \neq oa$, then one of the circles C_i, say C_1, contains a, so $C_1 \cap L = \{a', a''\}$, a contradiction. Hence the common tangent at o of $C_1, C_2, \dots, C_{q-1}$ is the line oa. Now we consider conjugate points $\overline{d}', \overline{d}''$ on L with $(a'a''\overline{d}'\overline{d}'') = -1$ and $\{d', d''\} \neq \{\overline{d}', \overline{d}''\}$. Again there arise $q-1$ circles $\overline{C}_1, \overline{C}_2, \dots, \overline{C}_{q-1}$ which are mutually tangent at o with common tangent oa. Hence $\{C_1, C_2, \dots, C_{q-1}\} = \{\overline{C}_1, \overline{C}_2, \dots, \overline{C}_{q-1}\}$, so $C_1 \cap L = \{d', d''\} = \{\overline{d}', \overline{d}''\}$, a contradiction. It follows that on $ob - \{o\}$ there is at least one affine point c such that the flock defined by $\{o, c\}$ is linear. Now by the conclusions of Parts 2 and 3 it is clear that for each point $d \in AG(2,q) - \{o\}$, the flock defined by $\{o, d\}$ has $\{a', a''\}$ as carriers. Let d', d'' be conjugate on L with $(a'a''d'd'') = -1$ and let C be the circle through o, b, d', d''. Clearly all points of $C - \{o\}$ define a flock of Thas type. Let $C - \{o\} = \{d_1, d_2, \dots, d_q\}$ and let $\overline{d}', \overline{d}''$ be conjugate on L with $(a'a''\overline{d}'\overline{d}'') = -1$ and $\{d', d''\} \neq \{\overline{d}', \overline{d}''\}$. Let D_i be the circle through $o, d_i, \overline{d}', \overline{d}''$, with $i = 1, 2, \dots, q$. The circles $D_1, D_2, \dots, D_q$ are mutually distinct and mutually tangent at o. So there arise $|D_1 \cup D_2 \cup \dots \cup D_q| = q^2+1$ affine points, a contradiction.

Next, suppose that o, a, b are collinear, with $\{o, a\}$ defining a linear flock and $\{o, b\}$ defining a Thas flock. Let $c \in AG(2,q) - oa$. By the previous section $\{o, c\}$ defines a linear flock. Now, again by the previous section, we see that $\{o, b\}$ defines a linear flock, a final contradiction.

Conclusion of this Part 5. Either all points of $AG(2,q) - \{o\}$ define a linear flock or all points of $AG(2,q) - \{o\}$ define a Thas flock.

Part 6. By the conclusions of Parts 4 and 5 all points of $AG(2,q) - \{o\}$ define a linear flock. By the conclusion of Part 2 the carriers defined by any point of $AG(2,q) - \{o\}$ are independent of the choice of that point. Let these common carriers be denoted by o', o''. Clearly each conical circle through o (i.e., each circle through o which is not a line of $AG(2,q)$) contains o' and o''.

Let $o_1 \in AG(2,q) - \{o\}$. Interchanging roles of o and o_1, the point o_1 defines carriers o_1' and o_1''. Considering a conical circle through o and o_1, it follows that $\{o', o''\} = \{o_1', o_1''\}$. Consequently each conical circle through o_1 also contains o' and o''.

It follows that any of the $q^3 - q^2$ conical circles contains o' and o''. Now by Part 2 $\mathcal{I}$ is the Miquelian inversive plane. ∎

The theorem can also be formulated as follows.

Theorem: *For $q \notin \{11, 23, 59\}$ the Desarguesian affine plane of odd order q has a unique extension. This extension is the Miquelian inversive plane of order q.*

Corollary: *Up to isomorphism there is a unique inversive plane of order 7.*

Proof: Let $\mathcal{I}$ be an inversive plane of order 7. For any point p the internal plane $\mathcal{I}_p$ is an affine plane of order 7. Since there is a unique plane of order 7 we have $\mathcal{I}_p = AG(2,7)$. Now by the theorem $\mathcal{I}$ is the Miquelian inversive plane of order 7. ∎

This provides a computer-free proof of the uniqueness of the inversive plane of order 7.

References:

1. L. Bader, Some new examples of flocks of $Q^+(3,q)$, *Geometriae Ded.*, to appear.
2. L. Bader and G. Lunardon, On the flocks of $Q^+(3,q)$. To appear.
3. R. D. Baker and G. L. Ebert, A nonlinear flock in the Minkowski plane of order 11. To appear.
4. Y. Chen, The Steiner system $S(3,6,26)$, *J. Geometry* 2 (1972), 7–28.
5. Y. Chen and G. Kaerlein, Eine Bemerkung über endliche Laguerre- und Minkowski-Ebenen, *Geometriae Ded.* 2 (1973), 193–194.
6. F. De Clerck, H. Gevaert, and J. A. Thas, Flocks of a quadratic cone in $PG(3,q)$, $q \leq 8$, *Geometriae Ded.* 26 (1988), 215–230.
7. P. Dembowski, *Finite Geometries*. Springer-Verlag, 1968.
8. R. H. F. Denniston, Uniqueness of the inversive plane of order 5, *Manuscr. Math.* 8 (1973), 11–19.
9. R. H. F. Denniston, Uniqueness of the inversive plane of order 7, *Manuscr. Math.* 8 (1973), 21–26.
10. J. C. Fisher and J. A. Thas, Flocks in $PG(3,q)$, *Math. Z.* 169 (1979), 1–11.
11. H. Gevaert and N. L. Johnson, Flocks of quadratic cones, generalized quadrangles and translation planes, *Geometriae Ded.* 27 (1988), 301–317.

12. H.-R. Halder and W. Heise, *Einführung in die Kombinatorik*. Carl Hanser Verlag, 1976.

13. J. W. P. Hirschfeld, *Projective Geometries over Finite Fields*. Clarendon Press, Oxford, 1979.

14. J. W. P. Hirschfeld, *Finite Projective Spaces of Three Dimensions*. Clarendon Press, Oxford, 1985.

15. N. L. Johnson, Flocks of hyperbolic quadrics and translation planes admitting affine homologies, *J. Geometry*, to appear.

16. M. Kallaher, Bol quasifields of dimension two over their kernels, *Arch. Math.* 25 (1974), 419–423.

17. M. Kallaher, On finite Bol quasifields, *Algebras, Groups and Geometry* 3 (1985), 300–312.

18. S. E. Payne and J. A. Thas, Generalized quadrangles with symmetry, Part II, *Simon Stevin* 49 (1976), 81–103.

19. J. A. Thas, Flocks of finite egglike inversive planes. In *Finite Geometric Structures and Their Applications* (ed. A. Barlotti), Cremonese Roma (1973), 189–191.

20. J. A. Thas, Flocks of nonsingular ruled quadrics in $PG(3,q)$, *Rend. Accad. Naz. Lincei* 59 (1975), 83–85.

21. J. A. Thas, Some results on quadrics and a new class of partial geometries, *Simon Stevin* 55 (1981), 129–139.

22. J. A. Thas, Generalized quadrangles and flocks of cones, *European J. Comb.* 8 (1987), 441–452.

23. J. A. Thas, Flocks, maximal exterior sets and inversive planes. In *Contemporary Mathematics (AMS)*, to appear.

24. E. Witt, Uber Steinersche Systeme, *Abh. Math. Sem. Univ. Hamb.* 12 (1938), 265–275.